60,-

Pitman Research Notes in Mathematics Series

D0775758

Submission of proposals for consideration
Suggestions for publication, in the form of outlines and representative samples, are invited by the Editorial Board for assessment. Intending authors should approach one of the main editors or another member of the Editorial Board, citing the relevant AMS subject classifications. Alternatively, outlines may be sent directly to the publisher's offices. Refereeing is by members of the board and other mathematical authorities in the topic concerned, throughout the world.

Preparation of accepted manuscripts
On acceptance of a proposal, the publisher will supply full instructions for the preparation of manuscripts in a form suitable for direct photo-lithographic reproduction. Specially printed grid sheets are provided and a contribution is offered by the publisher towards the cost of typing. Word processor output, subject to the publisher's approval, is also acceptable.

Illustrations should be prepared by the authors, ready for direct reproduction without further improvement. The use of hand-drawn symbols should be avoided wherever possible, in order to maintain maximum clarity of the text.

The publisher will be pleased to give any guidance necessary during the preparation of a typescript, and will be happy to answer any queries.

Important note
In order to avoid later retyping, intending authors are strongly urged not to begin final preparation of a typescript before receiving the publisher's guidelines and special paper. In this way it is hoped to preserve the uniform appearance of the series.

Longman Scientific & Technical
Longman House
Burnt Mill
Harlow, Essex, UK
(tel (0279) 26721)

Titles in this series

1 Improperly posed boundary value problems
A Carasso and A P Stone

2 Lie algebras generated by finite dimensional ideals
I N Stewart

3 Bifurcation problems in nonlinear elasticity
R W Dickey

4 Partial differential equations in the complex domain
D L Colton

5 Quasilinear hyperbolic systems and waves
A Jeffrey

6 Solution of boundary value problems by the method of integral operators
D L Colton

7 Taylor expansions and catastrophes
T Poston and I N Stewart

8 Function theoretic methods in differential equations
R P Gilbert and R J Weinacht

9 Differential topology with a view to applications
D R J Chillingworth

10 Characteristic classes of foliations
H V Pittie

11 Stochastic integration and generalized martingales
A U Kussmaul

12 Zeta-functions: An introduction to algebraic geometry
A D Thomas

13 Explicit *a priori* inequalities with applications to boundary value problems
V G Sigillito

14 Nonlinear diffusion
W E Fitzgibbon III and H F Walker

15 Unsolved problems concerning lattice points
J Hammer

16 Edge-colourings of graphs
S Fiorini and R J Wilson

17 Nonlinear analysis and mechanics: Heriot-Watt Symposium Volume I
R J Knops

18 Actions of fine abelian groups
C Kosniowski

19 Closed graph theorems and webbed spaces
M De Wilde

20 Singular perturbation techniques applied to integro-differential equations
H Grabmüller

21 Retarded functional differential equations: A global point of view
S E A Mohammed

22 Multiparameter spectral theory in Hilbert space
B D Sleeman

24 Mathematical modelling techniques
R Aris

25 Singular points of smooth mappings
C G Gibson

26 Nonlinear evolution equations solvable by the spectral transform
F Calogero

27 Nonlinear analysis and mechanics: Heriot-Watt Symposium Volume II
R J Knops

28 Constructive functional analysis
D S Bridges

29 Elongational flows: Aspects of the behaviour of model elasticoviscous fluids
C J S Petrie

30 Nonlinear analysis and mechanics: Heriot-Watt Symposium Volume III
R J Knops

31 Fractional calculus and integral transforms of generalized functions
A C McBride

32 Complex manifold techniques in theoretical physics
D E Lerner and P D Sommers

33 Hilbert's third problem: scissors congruence
C-H Sah

34 Graph theory and combinatorics
R J Wilson

35 The Tricomi equation with applications to the theory of plane transonic flow
A R Manwell

36 Abstract differential equations
S D Zaidman

37 Advances in twistor theory
L P Hughston and R S Ward

38 Operator theory and functional analysis
I Erdelyi

39 Nonlinear analysis and mechanics: Heriot-Watt Symposium Volume IV
R J Knops

40 Singular systems of differential equations
S L Campbell

41 N-dimensional crystallography
R L E Schwarzenberger

42 Nonlinear partial differential equations in physical problems
D Graffi

43 Shifts and periodicity for right invertible operators
D Przeworska-Rolewicz

44 Rings with chain conditions
A W Chatters and C R Hajarnavis

45 Moduli, deformations and classifications of compact complex manifolds
D Sundararaman

46 Nonlinear problems of analysis in geometry and mechanics
M Atteia, D Bancel and I Gumowski

47 Algorithmic methods in optimal control
W A Gruver and E Sachs

48 Abstract Cauchy problems and functional differential equations
F Kappel and W Schappacher

49 Sequence spaces
W H Ruckle

50 Recent contributions to nonlinear partial differential equations
H Berestycki and H Brezis

51 Subnormal operators
J B Conway

52 Wave propagation in viscoelastic media
F Mainardi

53 Nonlinear partial differential equations and their applications: Collège de France Seminar. Volume I
H Brezis and J L Lions

54 Geometry of Coxeter groups
H Hiller
55 Cusps of Gauss mappings
T Banchoff, T Gaffney and C McCrory
56 An approach to algebraic K-theory
A J Berrick
57 Convex analysis and optimization
J-P Aubin and R B Vintner
58 Convex analysis with applications in the differentiation of convex functions
J R Giles
59 Weak and variational methods for moving boundary problems
C M Elliott and J R Ockendon
60 Nonlinear partial differential equations and their applications: Collège de France Seminar. Volume II
H Brezis and J L Lions
61 Singular systems of differential equations II
S L Campbell
62 Rates of convergence in the central limit theorem
Peter Hall
63 Solution of differential equations by means of one-parameter groups
J M Hill
64 Hankel operators on Hilbert space
S C Power
65 Schrödinger-type operators with continuous spectra
M S P Eastham and H Kalf
66 Recent applications of generalized inverses
S L Campbell
67 Riesz and Fredholm theory in Banach algebra
B A Barnes, G J Murphy, M R F Smyth and T T West
68 Evolution equations and their applications
F Kappel and W Schappacher
69 Generalized solutions of Hamilton-Jacobi equations
P L Lions
70 Nonlinear partial differential equations and their applications: Collège de France Seminar. Volume III
H Brezis and J L Lions
71 Spectral theory and wave operators for the Schrödinger equation
A M Berthier
72 Approximation of Hilbert space operators I
D A Herrero
73 Vector valued Nevanlinna Theory
H J W Ziegler
74 Instability, nonexistence and weighted energy methods in fluid dynamics and related theories
B Straughan
75 Local bifurcation and symmetry
A Vanderbauwhede
76 Clifford analysis
F Brackx, R Delanghe and F Sommen
77 Nonlinear equivalence, reduction of PDEs to ODEs and fast convergent numerical methods
E E Rosinger
78 Free boundary problems, theory and applications. Volume I
A Fasano and M Primicerio
79 Free boundary problems, theory and applications. Volume II
A Fasano and M Primicerio
80 Symplectic geometry
A Crumeyrolle and J Grifone
81 An algorithmic analysis of a communication model with retransmission of flawed messages
D M Lucantoni
82 Geometric games and their applications
W H Ruckle
83 Additive groups of rings
S Feigelstock
84 Nonlinear partial differential equations and their applications: Collège de France Seminar. Volume IV
H Brezis and J L Lions
85 Multiplicative functionals on topological algebras
T Husain
86 Hamilton-Jacobi equations in Hilbert spaces
V Barbu and G Da Prato
87 Harmonic maps with symmetry, harmonic morphisms and deformations of metrics
P Baird
88 Similarity solutions of nonlinear partial differential equations
L Dresner
89 Contributions to nonlinear partial differential equations
C Bardos, A Damlamian, J I Díaz and J Hernández
90 Banach and Hilbert spaces of vector-valued functions
J Burbea and P Masani
91 Control and observation of neutral systems
D Salamon
92 Banach bundles, Banach modules and automorphisms of C^*-algebras
M J Dupré and R M Gillette
93 Nonlinear partial differential equations and their applications: Collège de France Seminar. Volume V
H Brezis and J L Lions
94 Computer algebra in applied mathematics: an introduction to MACSYMA
R H Rand
95 Advances in nonlinear waves. Volume I
L Debnath
96 FC-groups
M J Tomkinson
97 Topics in relaxation and ellipsoidal methods
M Akgül
98 Analogue of the group algebra for topological semigroups
H Dzinotyiweyi
99 Stochastic functional differential equations
S E A Mohammed
100 Optimal control of variational inequalities
V Barbu
101 Partial differential equations and dynamical systems
W E Fitzgibbon III
102 Approximation of Hilbert space operators. Volume II
C Apostol, L A Fialkow, D A Herrero and D Voiculescu
103 Nondiscrete induction and iterative processes
V Ptak and F-A Potra

104 Analytic functions – growth aspects
O P Juneja and G P Kapoor
105 Theory of Tikhonov regularization for Fredholm equations of the first kind
C W Groetsch
106 Nonlinear partial differential equations and free boundaries. Volume I
J I Díaz
107 Tight and taut immersions of manifolds
T E Cecil and P J Ryan
108 A layering method for viscous, incompressible L_p flows occupying R^n
A Douglis and E B Fabes
109 Nonlinear partial differential equations and their applications: Collège de France Seminar. Volume VI
H Brezis and J L Lions
110 Finite generalized quadrangles
S E Payne and J A Thas
111 Advances in nonlinear waves. Volume II
L Debnath
112 Topics in several complex variables
E Ramírez de Arellano and D Sundararaman
113 Differential equations, flow invariance and applications
N H Pavel
114 Geometrical combinatorics
F C Holroyd and R J Wilson
115 Generators of strongly continuous semigroups
J A van Casteren
116 Growth of algebras and Gelfand–Kirillov dimension
G R Krause and T H Lenagan
117 Theory of bases and cones
P K Kamthan and M Gupta
118 Linear groups and permutations
A R Camina and E A Whelan
119 General Wiener–Hopf factorization methods
F-O Speck
120 Free boundary problems: applications and theory, Volume III
A Bossavit, A Damlamian and M Fremond
121 Free boundary problems: applications and theory, Volume IV
A Bossavit, A Damlamian and M Fremond
122 Nonlinear partial differential equations and their applications: Collège de France Seminar. Volume VII
H Brezis and J L Lions
123 Geometric methods in operator algebras
H Araki and E G Effros
124 Infinite dimensional analysis–stochastic processes
S Albeverio
125 Ennio de Giorgi Colloquium
P Krée
126 Almost-periodic functions in abstract spaces
S Zaidman
127 Nonlinear variational problems
A Marino, L Modica, S Spagnolo and M Degiovanni
128 Second-order systems of partial differential equations in the plane
L K Hua, W Lin and C-Q Wu
129 Asymptotics of high-order ordinary differential equations
R B Paris and A D Wood
130 Stochastic differential equations
R Wu
131 Differential geometry
L A Cordero
132 Nonlinear differential equations
J K Hale and P Martinez-Amores
133 Approximation theory and applications
S P Singh
134 Near-rings and their links with groups
J D P Meldrum
135 Estimating eigenvalues with *a posteriori/a priori* inequalities
J R Kuttler and V G Sigillito
136 Regular semigroups as extensions
F J Pastijn and M Petrich
137 Representations of rank one Lie groups
D H Collingwood
138 Fractional calculus
G F Roach and A C McBride
139 Hamilton's principle in continuum mechanics
A Bedford
140 Numerical analysis
D F Griffiths and G A Watson

D F Griffiths & G A Watson (Editors)
University of Dundee

Numerical analysis

Longman Scientific & Technical
Copublished in the United States with
John Wiley & Sons, Inc., New York

Longman Scientific & Technical
Longman Group UK Limited
Longman House, Burnt Mill, Harlow
Essex CM20 2JE, England
and Associated Companies throughout the world.

Copublished in the United States with
John Wiley & Sons, Inc., 605 Third Avenue, New York, NY 10158

First published 1986

AMS Subject Classifications: (main) 65–06, 65D10, 65F99
(subsidiary) 65K10, 65L05, 65M25

ISSN 0269–3674

British Library Cataloguing in Publication Data
Numerical analysis.—(Pitman research notes in
mathematics series. ISSN 0269–3674; 140)
1. Numerical analysis
I. Griffiths, David F. II. Watson, G.A.
519.4 QA297

ISBN 0-582-98897-7

Library of Congress Cataloging-in-Publication Data
Numerical analysis.
(Pitman research notes in mathematics, ISSN 0269–3674; 140)
Proceedings of the 11th Dundee Biennial Conference on
Numerical Analysis held at the University of Dundee,
June 25–28, 1985.
Includes bibliographies.
1. Numerical analysis — Congresses. I. Griffiths,
D. F. (David Francis) II. Watson, G. A. III. Dundee
Conference on Numerical Analysis (11th: 1985)
IV. University of Dundee. V. Series.
QA297.N825 1986 519.4 86–7517
ISBN 0–470–20669–1 (USA only)

Reproduced and printed
in Great Britain by Biddles Ltd, Guildford and Kings Lynn.

Contents

Preface

P ALFELD: On the dimension of multivariate piecewise polynomials ... 1

F BREZZI: Theoretical and numerical problems in semi-conductor devices ... 24

G J COOPER: An algebraic condition for A-stable Runge-Kutta methods ... 32

S FRIEDLAND, J NOCEDAL and M L OVERTON: Four quadratically convergent methods for solving inverse eigenvalue problems ... 47

D GOLDFARB: Strategies for constraint deletion in active set algorithms ... 66

S-P HAN: Optimization by updated conjugate subspaces ... 82

A ISERLES: Order stars and stability barriers ... 98

C VAN LOAN: Parallel algorithms for constrained and unconstrained least squares problems ... 112

P S MARCUS: Numerical simulation of quasi-geostrophic flow using vortex and spectral methods ... 125

A R MITCHELL and D F GRIFFITHS: Beyond the linearised stability limit in non-linear problems ... 140

K W MORTON and A PRIESTLY: On characteristic Galerkin and Lagrange Galerkin methods ... 157

A H G RINNOOY KAN and G T TIMMER: The multi-level single linkage method for unconstrained and constrained global optimization ... 173

J M SANZ-SERNA and F VADILLO: Nonlinear instability, the dynamic approach ... 187

L N TREFETHEN: Dispersion, dissipation and stability ... 200

J G VERWER: Convergence and order reduction of diagonally implicit Runge-Kutta schemes in the method of lines ... 220

W L WENDLAND: Splines versus trigonometric polynomials - the h- versus the p-version in two-dimensional boundary integral methods ... 238

Submitted papers 256

Preface

The 11th Dundee Biennial Conference on Numerical Analysis was held in the University of Dundee during the 4 days 25-28 June, 1985. We were pleased to have more than 220 mathematicians in attendance, with over half of them coming from outside the U.K. The technical program was constructed around 16 invited talks, given by leading numerical analysts representative of a wide range of interests: full versions of their papers are published here. In addition to the invited papers, 80 shorter contributed talks were included in the program in three parallel sessions. Although this represented an increase over the number of talks given in previous conferences, time and space considerations meant that many potential speakers still had to be disappointed. The titles of all contributed talks given at the meeting, together with the addresses of the presenters, are listed here.

We would like to take this opportunity of thanking all the speakers, including the after dinner speaker at the conference dinner, Professor M.J.D. Powell, all chairmen and participants for their contributions. We are also grateful for the help received from many members of the Department of Mathematical Sciences of this University, both before and during the conference.

Financial support for the meeting was obtained from the European Research Office of the United States Army, and this support is gratefully acknowledged. The conference is also indebted to Dundee University for the provision of a sherry reception for all participants, and also for making available various University facilities throughout the week.

Regular subscribers to the Proceedings of the Dundee Numerical Analysis Conferences will have noted that this volume marks a change in the usual publishing arrangements. We wish to thank Pitman Publishing Limited for their cooperation in this venture, and in particular Bridget Buckley, Advanced Publishing Program Editor, for guiding us through the pre-publication stages.

October 1985

D.F. Griffiths
G.A. Watson

INVITED SPEAKERS

P Alfeld :	Department of Mathematics, University of Utah, Salt Lake City, Utah 84112, USA.
F Brezzi:	Dipartimento de Meccanica Strutturale dell' Universita e Istituto di Analisi Numerica del C.N.R, 27100 Pavia, Italy.
G J Cooper:	School of Mathematical and Physical Sciences, University of Sussex, Brighton, BN1 9QH, U.K.
D Goldfarb:	Department of Industrial Engineering and Operations Research, Columbia University, Seeley W. Mudd Building, New York, NY 10027, USA.
Shih-Ping Han:	Department of Mathematics, University of Illinois, Urbana, IL 61801, USA.
A Iserles:	Department of Applied Mathematics and Theoretical Physics, University of Cambridge, Silver Street, Cambridge, CB3 9EW, UK.
C van Loan:	Department of Computer Science, Cornell University, 405 Upson Hall, Ithaca, NY 14853, USA.
P Marcus:	Center for Astrophysics, Harvard University, 60 Garden Street, Cambridge, Massachusetts, MA02138, USA.
A R Mitchell:	Department of Mathematical Sciences, University of Dundee, Dundee, DD1 4HN, UK.
K W Morton:	Computing Laboratory, University of Oxford, 8-11 Keble Road, Oxford, OX1 3QD, UK.
M L Overton:	Courant Institute, New York University, 251 Mercer Street, New York, NY 10012, USA.
A H G Rinnooy Kan:	Erasmus Universteit Rotterdam, Postbus 1738, 3000 DR Rotterdam, The Netherlands.
J M Sanz-Serna:	Depto de Ecuaciones Functionales, Universidad de Valladolid, Facultad de Ciencias, Valladolid, Spain.
L N Trefethen:	Department of Mathematics, Mathematics Institute of Technology, Cambridge, Massachusetts, MA 02139, USA.
J G Verwer:	Mathematical Centre, Kruislaan 413, 1098 SJ Amsterdam, The Netherlands.
W L Wendland:	Technische Hochschule Darmstadt, Fachbereich Mathematik, Schlossgartenstrasse 7, 6100 Darmstadt, W. Germany.

P ALFELD

On the dimension of multivariate piecewise polynomials

1. INTRODUCTION

The importance of piecewise polynomial functions in numerical analysis and computer aided geometric design is obvious. Fundamental to any utilization of such functions is knowledge of the dimension, and, if possible, a basis of the given space. For functions of several variables this problem turns out to be surprisingly difficult and many questions are still unanswered.

Of main concern in this paper are spaces of once or twice differentiable piecewise polynomial functions that are defined on the tessellation of a simply connected polyhedral domain into tetrahedra. Lower bounds on the dimensions of such spaces are given. The bounds actually give the true dimension in all known cases with the exception of certain geometric degeneracies.

The paper is organized as follows: in Section 2, the trivial univariate case is described, to provide a contrast with the vastly more complicated multivariate case. In Section 3, our analysis techniques are illustrated by examining piecewise polynomial functions of two variables. The lower bounds so found agree with Schumaker's, 1979, but the proof technique is different. Section 4 constitutes the heart of the paper. Lower bounds on the dimension are given for spaces of piecewise polynomial functions defined on a tetrahedralization. The original motivation for this work was to use piecewise polynomial functions for the interpolation of scattered data. Thus this application is discussed briefly in Section 5.

Throughout this paper, we will use the convention that a binomial coefficient $\binom{n}{k} = 0$ if $n < k$.

Essential for an understanding of the proofs in this paper is familiarity with the Bézier-Bernstein form of multivariate polynomials. Continuous piecewise polynomial functions can be represented by their generalized Bézier nets where Bézier control points on edges joining two triangles (or faces joining two tetrahedra) have been identified. For an introduction to the Bézier form of a polynomial see Farin, 1980. Generalized Bézier ordinates are introduced in Alfeld, 1984.

The dimensions in all examples have been checked by direct calculation, using the symbol manipulation language REDUCE (see Hearn, 1983).

2. THE UNIVARIATE CASE

It is instructive to examine the univariate case as a backdrop for viewing the overwhelming complexity of functions of several variables. Suppose we are given a partition of the interval $[a,b]$ by a set of intervals $[x_{n-1},x_n]$ where $n = 1,2,\ldots,N$ and $a = x_0 < x_1 < \ldots < x_N = b$. The space of interest is then

$$S_1^{k,m} := \{p \in C^k[a,b] : p \text{ is a polynomial of degree } m \text{ on each } [x_{n-1},x_n]\}.$$

In order for this space to be nontrivial k must be $< m$. Then,

$$\dim(S_1^{k,m}) = m + 1 + (N-1)(m-k)$$

as can be easily verified: Let $p \in S_1^{k,m}$. On the first interval, $[x_0,x_1]$, p may be any polynomial, yielding $m+1$ degrees of freedom. On each of the subsequent $N-1$ intervals, $k+1$ degrees of freedom are needed to ensure a C^1 joint with the previous interval, leaving $m-k$ degrees of freedom. (The corresponding interpolation problem can be solved uniquely, see Davis, 1975, p.28).

3. THE BIVARIATE CASE

3.1 The Geometry of Triangulations

We consider a polygonal simply connected domain D_2 which has been triangulated by triangles Δ_i, $i = 1,\ldots,T$, and denote the triangulation by $\mathcal{T}(D_2)$. In the bivariate case, it is convenient to express all relevant quantities in terms of the number of boundary vertices (B) and interior vertices (I) of $\mathcal{T}(D_2)$. The number of triangles, T, is given by

$$T = B + 2I - 2 \tag{1}$$

and the number of interior edges, E, satisfies

$$E = B + 3I - 3 \tag{2}$$

(see Ewing et al, 1970). The number of boundary edges equals the number of boundary vertices.

We are concerned with the spaces

$$S_2^{k,m} := \{p \in C^k(D_2),\ p \text{ is polynomial of degree } m \text{ on each } \Delta_i\},$$

particularly for $k \in \{1,2\}$. In studying $S_2^{k,m}$ we will also use the spaces

$$P_2^{k,m} := \{p \in S_2^{0,m},\ p \text{ is } k \text{ times differentiable at all vertices of } \mathcal{T}(D_2)\}.$$

Obviously,

$$S_2^{k,m} \subset P_2^{k,m} . \tag{3}$$

The basic approach to obtaining lower bounds on the dimension of $S_2^{k,m}$ consists of embedding it into the larger but simpler space $P_2^{k,m}$ and then subtracting a number of sufficient conditions on functions in $P_2^{k,m}$ that force the more restrictive smoothness of a function in $S_2^{k,m}$.

Theorem 1. Let $m \geq 2k + 1$. Then

$$\dim(P_2^{k,m}) = \alpha B + \beta I + \gamma \tag{4}$$

where

$$\alpha = (-2k^2 - 2k + m^2 + m)/2$$

$$\beta = (-5k^2 - 3k + 2m^2)/2$$

$$\gamma = 3k^2 + 3k - m^2 + 1.$$

Proof. The key to the structure of $P_2^{k,m}$ is the fact that it can be easily parameterized. This can be best visualized by contemplating the generalized Bézier net of the piecewise polynomial function. To force k-th order differentiability at the vertices of $\mathcal{T}(D_2)$, all Bézier ordinates up to k layers distant from the vertices are determined by the $\binom{k+2}{2}$ 0-th through k-th order derivatives at the vertices. Since $m \geq 2k+1$, these derivatives can be specified independently and arbitrarily. Along each edge of the triangulation there are m+1 generalized Bézier ordinates of which 2(k+1) have been specified at the vertices, leaving m + 1 - 2(k+1) free parameters. In the interior of each triangle, there are $\binom{m-1}{2}$ Bézier ordinates, of which $\binom{k}{2}$ have been specified at each vertex, leaving $\binom{m-1}{2} - 3\binom{k}{2}$ degrees of freedom in the interior of each triangle. The derivatives at the vertices, as well as the Bézier ordinates in the interior of edges and triangles, can be specified independently and arbitrarily, and completely determine a function in $P_2^{k,m}$. Therefore,

$$\dim(P_2^{k,m}) = \binom{k+2}{2}(B+I) + (m-1-2k)(B+E) + \left[\binom{m-1}{2} - 3\binom{k}{2}\right]T$$

which, by (1) and (2) yields (4). □

For convenience, we list values of α, β, and γ for some values of m and k in Table 1.

Table 1: Dimension of $P_2^{k,m}$

m	k	α	β	γ
1	0	1	1	0
2	0	3	4	-3
3	0	6	9	-8
4	0	10	16	-15
3	1	4	5	-2
4	1	8	12	-9
5	1	13	21	-18
6	1	19	32	-29
5	2	9	12	-6
6	2	15	23	-17
7	2	22	36	-30
8	2	30	51	-45

3.2 The C^1 Case

One of the most bewildering features of spaces of piecewise polynomial functions is the fact that their dimension may depend not just on the *topology* but also on the precise *geometry* of the triangulation $T(D_2)$. Of particular interest in the bivariate C^1 case are singular vertices whose significance was first recognized by Powell, 1973. The following definition is due to Morgan and Scott, 1975a.

Definition 1. An interior vertex of $\mathcal{T}(D_2)$ for which the adjacent edges of each edge are collinear is called *singular*.

Thus the union of the four triangles sharing a singular vertex simply is a quadrilateral with the diagonals drawn in.

Theorem 2. Let $m \geq 3$. Then

$$\dim(S_2^{1,m}) \geq \alpha_1 B + \beta_1 I + \gamma_1 + \sigma \tag{5}$$

where

$$\alpha_1 = m(m-1)/2$$

$$\beta_1 = m^2 - 3m + 2$$

$$\gamma_1 = -m^2 + 3m + 1$$

and σ is the number of singular vertices.

Proof. Because of (3) a lower bound on $\dim(S_2^{1,m})$ can be obtained by subtracting from $\dim(P_2^{1,m})$ a number of conditions that force a function in $P_2^{1,m}$ to be in $S_2^{1,m}$. Consider a first order derivative of some $p \in P_2^{1,m}$, across any interior edge of $\mathcal{T}(D_2)$. On each of the two triangles sharing the edge, this derivative will be a polynomial of degree m-1 with m degrees of freedom. Since p is differentiable at the vertices of the edge, differentiability across the edge imposes m-2 additional conditions. Thus

$$\dim(S_2^{1,m}) \geq \dim(P_2^{1,m}) - (m-2)E$$

which together with (2) establishes (5) except for the term σ.

Now consider a singular vertex V_5, say and let V_1, V_2, V_3, and V_4 denote the vertices of the surrounding quadrilateral labeled in sequence. One cross-edge smoothness condition is implied by the smoothness conditions across the other three edges. To see this, assume that differentiability has been imposed across edges e_{25}, e_{35}, and e_{45} (where $e_{ij} := V_j - V_i$). Denote the restriction of p to the triangle with vertices V_i, V_j, and V_k by p_{ijk}. Then

$$\frac{\partial^2 p_{145}}{\partial e_{15}\partial e_{45}}(v_5) = \frac{\partial^2 p_{345}}{\partial e_{15}\partial e_{45}}(v_5) = \frac{\partial^2 p_{245}}{\partial e_{15}\partial e_{45}}(v_5) = \frac{\partial^2 p_{125}}{\partial e_{15}\partial e_{45}}(v_5).$$

Consider the first equality (across edge e_{45}). The mixed derivative is tangential (i.e., in the direction of edge e_{45}) of a first order derivative (in the direction of e_{15}) across edge e_{45}. That cross derivative is differentiable on both triangles sharing e_{45} since it is polynomial. It is also well defined along edge e_{45} since we assume that differentiability across edge e_{45} has been enforced. Thus the first of the above equalities is valid. The second equality follows similarly since the mixed derivative is tangential in the direction of e_{15} *which is a multiple of* e_{35}! The third equality follows in the same way. Thus a tangential derivative of a cross boundary derivative across e_{15} is continuous at the centroid, implying that one less condition has to be imposed on e_{15}. We obtain one such additional free condition at each singular vertex of $\mathcal{T}(D_2)$, which establishes (5). □

Remark 1. The lower bound (5) is identical to that obtained by Schumaker, 1979. Very early, Strang, 1973 and 1974, conjectured that the right hand side of (5) gives the dimension of $S_2^{1,m}$.

Remark 2. The significance of singular vertices is that they provide the simplest instance where the dimension of the piecewise polynomial space depends upon the geometry rather than just the topology of the underlying triangulation. If a singular vertex is present, then the additional degrees of freedom can be removed by an arbitrarily small perturbation of the points in the triangulation, without altering its topology. For $m \geq 5$ it has been shown by Morgan and Scott, 1975a, that singular vertices are the only possible source of geometric degeneracies. For $m \in \{3,4\}$, no degeneracies other than singular vertices are known, but efforts to disprove the existence of others have been unsuccessful. For $m = 2$, Morgan and Scott, 1975b, give a different example where the dimension depends upon the geometry. Their domain is triangular and there are three interior points forming an inverted triangle. That domain is triangulated by seven triangles. All *known* examples share the property that there exists a small perturbation of the underlying point set which will restore the dimension to the value given by the lower bound. It is not known if such a generic dimension exists for all triangulations.

Remark 3. The technique of reducing smoothness conditions to continuity conditions on high order mixed partial derivatives will be used extensively in the trivariate case. Notice that this technique divorces the question of smoothness from the particular polynomial structure of the interpolation space.

Table 2: Dimension of $S_2^{1,m}$

m	α_1	β_1	γ_1
3	3	2	1
4	6	6	-3
5	10	12	-9
6	15	20	-17

For convenience, Table 2 lists values of α_1, β_1, and γ_1, for small values of m.

3.3 The C^2 Case

The development in this subsection is analogous to that in the preceding one, except that an additional source of free smoothness conditions has to be taken into account.

Definition 2. An interior vertex of $\mathcal{T}(D_2)$ from which precisely three edges of different slopes emanate is called a *Clough-Tocher vertex*.

Example 1 (cf. Alfeld, 1985). Consider a triangle on which the Clough-Tocher split has been applied twice. Thus the triangle with vertices V_1, V_2, and V_3 is divided about its centroid $V_4 = (V_1+V_2+V_3)/3$. Each subtriangle is divided about its centroid $V_5 = (V_4+V_2+V_3)/3$, $V_6 = (V_1+V_4+V_3)/3$, or $V_7 = (V_1+V_2+V_4)/3$. Obviously, the subcentroids V_5, V_6, and V_7 are each a Clough-Tocher vertex. In addition, it can be seen easily that the triples $\{V_1,V_4,V_5\}$, $\{V_2,V_4,V_6\}$, and $\{V_3,V_4,V_7\}$ are each collinear. Thus the centroid V_4 is also a Clough-Tocher vertex.

Theorem 3. Let $m \geq 5$. Then

$$\dim(S_2^{2,m}) \geq \alpha_2 B + \beta_2 I + \gamma_2 + 3\sigma + \eta \tag{6}$$

where

$$\alpha_2 = (m^2-3m+2)/2$$

$$\beta_2 = m^2 - 6m + 8$$

$$\gamma_2 = -m^2 + 6m - 2,$$

σ is the number of singular vertices, and η is the number of Clough-Tocher vertices.

Proof. The proof is similar to that of Theorem 2. On each interior edge we impose m-4 conditions on the first derivative, and m-3 conditions on the second derivative. Subtracting this number from $\dim(P_2^{2,m})$ yields (6) except for the σ and η terms. At each singular vertex, three free conditions are obtained. This follows as for Theorem 2 by considering the derivatives

$$\frac{\partial^3 p}{\partial e_{15}^2 \partial e_{25}} (V_5),$$

$$\frac{\partial^3 p}{\partial e_{15} \partial e_{25}^2} (V_5),$$

and

$$\frac{\partial^4 p}{\partial e_{15}^2 \partial e_{25}^2} (V_5) .$$

Now consider a Clough-Tocher vertex V_4 with V_1, V_2, and V_3 being the end-points of three edges with different slopes emanating from V_4. Then the derivative

$$\frac{\partial^3 p}{\partial e_{14} \partial e_{24} \partial e_{34}}$$

will be continuous at V_4 if it is continuous across just two of the three edges. Continuity across any additional parallel edge is implied by continuity across its counterpart in $\{e_{14}, e_{24}, e_{34}\}$. □

Remark. The lower bound (6) is identical to that obtained by Schumaker, 1979. For $m \geq 5$ there is no known case in which the actual dimension of $S_2^{2,m}$ exceeds the lower bound given in (6).

Table 3: Dimension of $S_2^{2,m}$

m	α_2	β_2	γ_2
5	6	3	3
6	10	8	-2
7	15	15	-9
8	21	24	-18

For convenience, Table 3 lists values of α_2, β_2, and γ_2, for small values of m.

Example 2. Consider the double Clough-Tocher split in Example 1. There, $B = 3$, $I = 4$, $\sigma = 0$, and $\eta = 4$. Then

$$\dim(S_2^{2,5}) = 37.$$

If the location of the centroid V_4 is perturbed slightly, then η becomes 3, and

$$\dim(S_2^{2,5}) = 36.$$

Further perturbations of the centroid, or the subcentroids, do not alter the dimension.

4. THE TRIVARIATE CASE

4.1 The Geometry of Tetrahedral Tessellations

We consider a polyhedral simply connected domain D_3 that has been tessellated into tetrahedra T_i, $i = 1,\dots,T$, and denote the tessellation by $\mathcal{T}(D_3)$.

In the trivariate case, it is impossible to express topological quantities like the number of tetrahedra in terms of the number of (interior and boundary) vertices of the domain. Indeed, the number of tetrahedra can grow faster than any linear function of the number of vertices.

Example 3. Consider a tetrahedral tessellation obtained from a triangulated convex cell with no flaps (Schumaker, 1979), having μ boundary vertices and one interior vertex, and containing μ triangles. A tessellation into μ tetrahedra is built by adding a point above the cell and tessellating the resulting pyramid. Arbitrarily many additional layers are then generated by repeatedly adding a point above the current pyramid and tessellating the new layer. If ν is the number of layers, then the tessellation consists of $\mu+2$ boundary vertices, $\nu-1$ interior vertices, and $\mu\nu$ tetrahedra. Since μ can be arbitrarily large, the number of tetrahedra can grow arbitrarily much faster than the number of vertices.

In expressing topological quantities we require a substitute for the numbers of interior and boundary vertices. To obtain this we consider the tessellation $T(D_3)$ to have been built by adding one tetrahedron at a time such that each new tetrahedron joins the current tessellation at 1,2, or 3 faces, and no additional edges or vertices. We denote by a_i the *number of times* that a new tetrahedron was joined at *precisely* i *faces* (i = 1,2,3).

All relevant topological quantities can now be expressed in terms of the a_i. Of particular significance is the number of *shared edges*, i.e., those edges that the new tetrahedron shares with the current tessellation, but that do not become interior edges when the new tetrahedron is joined. Table 4 gives the initial values (i.v.) of the number of interior vertices (I), interior and boundary vertices (V), interior edges (E), shared edges (S), interior faces (F), and tetrahedra (T), and the changes in their values when a new tetrahedron is joined at i faces.

Table 4: Construction of Tetrahedral Tessellation

	i	I	V	E	S	F	T
i.v.:		0	4	0	0	0	1
	1	+0	+1	+0	+3	+1	+1
	2	+0	+0	+1	+4	+2	+1
	3	+1	+0	+3	+3	+3	+1

It follows from Table 4 that

$$a_1 = V-4, \quad a_3 = I, \quad a_2 = T-1-a_1-a_3.$$

The a_i, $i = 1,2,3$, are thus functions of the tessellation and independent of the sequence in which it is assembled.

Note that although a more convenient approach was available in the bivariate case we could have proceeded similarly as in the trivariate one. Then a_i, $(i = 1,2)$ would have denoted the number of times we join the new triangle on precisely i edges, and we would have had $a_1 = B+I-3$ and $a_2 = I$.

Example 4. In the construction described in Example 3, $a_1 = \mu+\nu-3$, $a_2 = \mu\nu-\mu-2\nu+3$, and $a_3 = \nu-1$.

Similarly as in the bivariate case, we consider the spaces

$$S_3^{k,m} := \{p \in C^k(D_3),\ p \text{ is polynomial of degree } m \text{ on each } T_i\},$$

and

$$P_3^{k,m} := \{p \in S_3^{0,m},\ p \text{ is } k \text{ times differentiable at all vertices of } \mathcal{T}(D_3)\}.$$

Theorem 4. Let $m \geq 2k+1$. Then

$$\dim(P_3^{k,m}) = \rho_0 + \sum_{i=1}^{3} \rho_i a_i \tag{7}$$

where

$$\rho_0 = \binom{m+3}{3}$$

$$\rho_1 = \binom{m+2}{3} - 3\binom{k+2}{3}$$

$$\rho_2 = \binom{m+1}{3} - 2\binom{k+1}{3} - 2\binom{k+2}{3}$$

$$\rho_3 = \binom{m}{3} - \binom{k}{3} - 3\binom{k+1}{3}.$$

Proof. The proof is similar to that of Theorem 1. Observe that the number of parameters in a trivariate polynomial of degree m is $\binom{m+3}{3}$. These parameters are best visualized as points in the (tetrahedral) Bézier net. Removing i faces of that net leaves $\binom{m+3-i}{3}$ Bézier control points. Similarly, derivatives through k-th order determine the $\binom{k+3}{3}$ Bézier control points in the first k layers next to each vertex. Stripping i layers leaves $\binom{k+3-i}{3}$ Bézier control points at the vertex.

The term ρ_0 accounts for the degrees of freedom on the initial tetrahedron.

When joining a tetrahedron on precisely one face, then the corresponding layer of Bézier control points is already present in the current tessellatic and contributes no free parameters. This accounts for the term $\binom{m+2}{3}$ in ρ_1. Similarly, the derivative parameters at three of the vertices of the new tetrahedron have already been disposed of. Of these, those lying in the new interior face have already been accounted for, leaving $\binom{k+2}{3}$ additional unavailable parameters at each of the three vertices defining the new interior face. Thus we obtain the factor ρ_1 multiplying a_1.

The factors ρ_2 and ρ_3 are obtained similarly. The term ρ_2 follows from the observation that two vertices of the new tetrahedron are common to the two new interior faces, resulting in the correction term $2\binom{k+1}{3}$, and two are contained in only one new interior face, resulting in the term $2\binom{k+2}{3}$. For ρ_3, the new interior vertex is shared by all three new interior faces, whereas the other three vertices are each shared by precisely two new interior faces. □

Table 5 lists the values of ρ_i, $i = 0,\ldots,3$, for some values of m and k.

Table 5: Dimension of $P_3^{k,m}$

m	k	ρ_0	ρ_1	ρ_2	ρ_3
1	0	4	1	0	0
2	0	10	4	1	0
3	0	20	10	4	1
4	0	35	20	10	4
5	0	56	35	20	10
3	1	20	7	2	1
4	1	35	17	8	4
5	1	56	32	18	10
6	1	84	53	33	20
7	1	120	81	54	35
5	2	56	23	10	7
6	2	84	44	25	17
7	2	120	72	46	32
8	2	165	108	74	53
9	2	220	153	110	81

4.2 The C^1 case

We have to take into account the equivalent of singular vertices:

Definition 3. An interior edge of $T(D_3)$ is called a *singular edge* if a section across it and the tetrahedra sharing it defines a singular vertex.

Theorem 5. Let $m \geq 3$. Then

$$\dim(S_3^{1,m}) \geq \dim(P_3^{1,m}) - (m-2)S - \binom{m-2}{2}F + (m-1)\sigma \tag{8}$$

where $S = 3a_1+4a_2+3a_3$, $F = a_1+2a_2+3a_3$, and σ is the number of singular edges.

Proof. Similarly as in the proof of Theorem 2, a lower bound on $\dim(S_3^{1,m})$ can be obtained by subtracting from $\dim(P_3^{1,m})$ a number of conditions that will force the more restrictive smoothness of a function in $S_3^{1,m}$. Since functions in $P_3^{1,m}$ are already differentiable at the vertices of $T(D_3)$, we need to consider only edges and faces that are shared by more than one tetrahedron. Let $p \in P_3^{1,m}$.

As in the bivariate case, m-2 conditions on p will force differentiability across any edge shared by two tetrahedra. However, differentiability across any *new interior edge* when joining a new tetrahedron to the current tessellation, is automatic. To see this consider a section across an interior edge shared by n tetrahedra, say, and assume that the tetrahedron corresponding to the sides a and b is being joined to the tessellation (see Figure 1). Let P be the point corresponding to the intersected edge, and denote by p_i, $i = 1,\ldots,n$ the restrictions of p to the i-th tetrahedron. Consider the derivative across the edge in the direction of b, and assume that differentiability has been enforced across the edge among all tetrahedra, except the newly joined one. Then

$$\frac{\partial p_n}{\partial b}(P) = \frac{\partial p_{n-1}}{\partial b}(P) = \cdots = \frac{\partial p_2}{\partial b}(P) = \frac{\partial p_1}{\partial b}(P).$$

Of these equalities, the first (across b) is valid since the partial derivative is tangential in the direction of b. All others are valid because differentiability has been enforced between the relevant tetrahedra. Together, the equalities imply continuity of $\partial p/\partial b$, and hence of any other directional derivative, across edge a. The continuity of $\partial p/\partial a$ across edge b follows similarly. The argument in this paragraph accounts for the term $(m-2)S$ in (8).

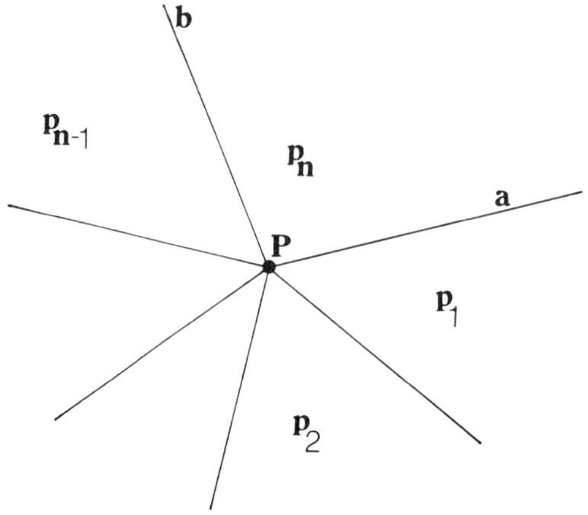

Figure 1: Section across an interior edge

Now consider a first order derivative of p across an interior face, and assume that differentiability has been enforced across its edges. The difference in the derivatives can be represented as a bivariate polynomial of degrees m-1 with $\binom{m+1}{2}$ parameters. Of these, those on the edges have been forced to be zero, leaving $\binom{m-2}{2}$ parameters that are required to be zero. This yields the term $\binom{m-2}{2}F$ in (8) (recall that F is the number of interior faces).

To account for the last term in (8), consider a section across a singular edge. As in the bivariate case, continuity of a mixed second order derivative restricted to the edge will be implied by continuity of first order derivatives across the other three faces. That derivative can be represented as a univariate polynomial of degree m-2 along the singular edge, having m-1 parameters, which are freed by the singularity. □

Corollary. Let $m \geq 3$. Then

$$\dim(S_3^{1,m}) \geq \max\{\tau_0, \tau_0 + \sum_{i=1}^{3} \tau_i a_i + (m-1)\sigma\} \tag{9}$$

where

$$\tau_0 = \rho_0 = (m^3+6m^2+11m+6)/6$$

$$\tau_1 = (m(m^2-1))/6$$

$$\tau_2 = (m(m^2-6m+5))/6$$

$$\tau_3 = (m^3-12m^2+29m-18)/6.$$

Proof. The first expression on the right hand side follows because the dimension of $\dim(S_3^{1,m})$ cannot be less than the dimension of the space of all polynomials of degree up to m. The second expression follows from (8) by direct calculation. □

Table 6 lists values of the τ_i for some values of m.

Table 6: Dimension of $S_3^{1,m}$

m	k	τ_0	τ_1	τ_2	τ_3
3	1	20	4	-2	-2
4	1	35	10	-2	-5
5	1	56	20	0	-8
6	1	84	35	5	-10
7	1	120	56	14	-10

Direct calculation shows that the right hand side of (9) gives the true value of $\dim(S_3^{1,m})$ for the following examples:

Example 5 (Clough-Tocher Split). Let the domain D_3 be a tetrahedron that has been split about its centroid and tessellated by four tetrahedra. Thus $a_1 = a_2 = a_3 = 1$. Then

$$\dim(S_3^{1,3}) = 20, \quad \dim(S_3^{1,4}) = 38, \quad \dim(S_3^{1,5}) = 68.$$

Example 6 (Four Tetrahedra). The domain is obtained as in Example 3, with $\mu = 4$ and $\nu = 1$, such that the interior edge is not singular. Thus $a_1 = 2$, $a_2 = 1$, $a_3 = 0$. Then

$$\dim(S_3^{1,3}) = 26 \quad \text{and} \quad \dim(S_3^{1,4}) = 53.$$

Example 7 (Farin's split). The domain is a tetrahedron split about the centroid, where additionally each subtetrahedron has been split about the centroid of its unique boundary face. Thus $a_1 = a_2 = 5$, $a_3 = 1$. Then

$$\dim(S_3^{1,3}) = 28 \quad \text{and} \quad \dim(S_3^{1,4}) = 70.$$

Farin has constructed* a trivariate piecewise cubic C^1 and scheme for C^1 data on that split.

Example 8 (Singular Edge). As Example 6, except that the interior edge is singular. Then

$$\dim(S_3^{1,3}) = 28 \quad \text{and} \quad \dim(S_3^{1,4}) = 56.$$

There are no cases in which the exact dimension of $S_3^{1,m}$ is known and does not agree with the right hand side of (9), except for some geometric degeneracies that will be reported elsewhere.

4.3 The C^2 case

We proceed by analogy to the C^1 case.

Definition 4. An interior edge of $\mathcal{T}(D_3)$ is called a *Clough-Tocher edge* if a section across it and the tetrahedra sharing it defines a Clough-Tocher vertex.

Theorem 6. Let $m \geq 5$. Then

$$\dim(S_3^{2,m}) \geq \dim(P_3^{2,m}) - \kappa_1 S - \kappa_2 F + \kappa_3 a_2 + \kappa_4 a_3 + \kappa\sigma + \chi_{21} a_{21} + \sum_{i=1}^{3} \chi_{3i} a_{3i} \qquad (10)$$

where

$$S = 3a_1 + 4a_2 + 3a_3, \quad F = a_1 + 2a_2 + 3a_3,$$

σ is the number of singular edges, a_{21} is the number of times that an a_2 step created a Clough-Tocher edge, a_{3i} is the number of times that an a_3 step created precisely i Clough-Tocher edges ($i \in \{1,2,3\}$), and

$$\kappa_1 = 2m-7,$$

$$\kappa_2 = \binom{m-2}{2} + \binom{m-3}{2},$$

---

* Private communication by Gerald Farin.

$\kappa_3 = m-3,$

$\kappa_4 = 3m-10,$

$\kappa = 3m-7,$

$\chi_{21} = m-2,$

$\chi_{31} = m-3, \quad \chi_{32} = 2(m-3), \quad \chi_{33} = 3m-10.$

Proof. We discuss the individual terms in turn:

κ_1: This follows as in the C^1 case. Only shared edges need to be considered. On each shared edge, m-4 C^1 and m-3 C^2 conditions are imposed, yielding a total of 2m-7 conditions.

κ_2: This also follows as in the C^1 case. A first order cross derivative on an interior face can be represented as a bivariate polynomial of degree m-1 with $\binom{m+1}{2}$ parameters, leaving $\binom{m-2}{2}$ parameters after stripping off the three edge layers. Similarly, considering second order cross derivatives yields $\binom{m-3}{2}$ conditions.

The conditions imposed in the above two items are sufficient, but not necessary to enforce second order differentiability. The following terms correct some of the discrepancy.

κ_3: Consider joining a new tetrahedron at precisely two faces F and G sharing the interior edge, e, say. Assume that second order differentiability along e has been enforced across all faces sharing e except face F. Let f be a direction in F and let n be a direction across F. Then the derivative $\mathcal{D}p := \left.\frac{\partial^2 p}{\partial f \partial n}\right|_e$ will be continuous across face F. $\mathcal{D}p$ can be represented as a univariate polynomial of degree m-2 with m-1 parameters. However, second order derivatives are already assumed to be continuous at the vertices so that only m-3 parameters are freed.

κ_4: This term follows similarly as the preceding one. Denote the three new interior faces by F, G, and H, and let n be a direction not contained in any of these faces. Let $e_1 = H \cap F$, $e_2 = F \cap G$, and

$e_3 = G \cap H$. Assume that second order differentiability has been enforced across all faces except F and G. Then the derivative $\left.\frac{\partial^2 p}{\partial e_2 \partial n}\right|_{e_3}$ will be continuous across face G, and $\left.\frac{\partial^2 p}{\partial e_2 \partial n}\right|_{e_1}$ will be continuous across face F, freeing 2(m-3) parameters. Now assume that second order differentiability has also been enforced across face F. This implies as before continuity of $\left.\frac{\partial^2 p}{\partial e_3 \partial n}\right|_{e_2}$. However, this last step frees only m-4 parameters since the continuity of $\frac{\partial^3 p}{\partial e_1 \partial e_2 \partial e_3}$ at the new interior vertex follows from the second order smoothness across edges e_1 and e_3. Thus we obtain a total of 3m-10 additional degrees of freedom.

κ: As in the bivariate case, two third order derivatives, and one fourth order derivative are continuous along the singular edge, freeing 2(m-2) + (m-3) = 3m-7 parameters.

χ_{21}: Let e_i, i = 1,2,3, be contained in the i-th of the faces sharing the Clough-Tocher edge e, and assume that second order differentiability has been enforced across two of the faces. Then, as in the bivariate case, the derivative $\frac{\partial^3 p}{\partial e_1 \partial e_2 \partial e_3}$ across the third face will be continuous along e, freeing m-2 parameters.

χ_{3i}: We use the same notation as in the discussion of the κ_4 term. First consider the term multiplied by χ_{31}. Assume e_1 is the Clough-Tocher edge and second order differentiability has been enforced across face H. Let f be a direction in F, h a direction in H, and n a direction in the third face sharing e_1. Then the derivative $\frac{\partial^3 p}{\partial f \partial h \partial n}$ is already continuous at the new interior vertex. Hence only m-3 parameters are freed, not m-2 as in the discussion of the χ_{21} term. The terms multiplied by χ_{32} and χ_{33} are obtained similarly, except that for χ_{33} second order differentiability across the edges e_1 and e_2 already frees one of the parameters that would be freed if e_3 was the only Clough-Tocher edge. □

Corollary. Let $m \geq 5$. Then

$$\dim(S_2^{2,m}) \geq \max\{\omega_0,\ \omega_0 + \sum_{i=1}^{3} \omega_i a_i + \kappa\sigma + \chi_{21}a_{21} + \sum_{i=1}^{3} \chi_{3i}a_{3i}\} \tag{11}$$

where

$$\omega_0 = \rho_0 = (m^3+6m^2+11m+6)/6$$

$$\omega_1 = (m(m^2-3m+2))/6$$

$$\omega_2 = (m^3-12m^2+29m-18)/6$$

$$\omega_3 = (m^3-21m^2+92m-114)/6.$$

Proof: This follows as in the proof of (9) and by direct calculation from (10). □

Table 7 lists the values of the constants in (11) for some values of m.

Table 7: Dimension of $S_3^{2,m}$

m	ω_0	ω_1	ω_2	ω_3	κ	χ_{21}	χ_{31}	χ_{32}	χ_{33}
5	56	10	-8	-9	8	3	2	4	5
6	84	20	-10	-17	11	4	3	6	8
7	120	35	-10	-26	14	5	4	8	11
8	165	56	-7	-35	17	6	5	10	14
9	220	84	0	-43	20	7	6	12	17
10	286	120	12	-49	23	8	7	14	20

Direct calculation shows that the right hand side of (11) gives the true value of $\dim(S_3^{2,m})$ for the following examples:

Example 9 (Clough-Tocher Split). The domain is as in Example 5, $a_{21} = a_{33} = 1$, and $a_{31} = a_{32} = \sigma = 0$. Then

$$\dim(S_3^{2,5}) = 57 \quad \text{and} \quad \dim(S_3^{2,6}) = 89.$$

Example 10 (Four Tetrahedra). The domain is as in Example 6, $a_3 = a_{21} = a_{31} = a_{32} = a_{33} = \sigma = 0$. Then

$$\dim(S_3^{2,5}) = 68 \quad \text{and} \quad \dim(S_3^{2,6}) = 114.$$

Example 11 (Singular Edge). The domain is as in Example 10, except that the interior edge is singular and $\sigma = 1$. Then

$$\dim(S_3^{2,5}) = 76 \quad \text{and} \quad \dim(S_3^{2,6}) = 125.$$

There are no cases in which the exact dimension of $S_3^{2,m}$ is known and does not agree with the right hand side of (11).

5. REMARKS ON INTERPOLATION BY PIECEWISE POLYNOMIALS

The work reported in this paper was motivated originally by the desire to use piecewise polynomials with globally enforced smoothness properties for the interpolation of scattered multivariate data. Most existing interpolation schemes defined on triangulations or tetrahedralizations enforce smoothness locally. See the recent survey by Barnhill, 1985, for more information. An exception to the general rule is described in Schmidt, 1982. The price to be paid for a local scheme is an increased local complexity. For example, to obtain a local bivariate piecewise polynomial C^1 scheme one either has to use a function that is *piecewise cubic* on each triangle (the Clough-Tocher scheme, see Strang and Fix, 1973, p.83, or Alfeld, 1984) or a function that is *quintic* on each triangle (see Strang and Fix, 1973, p.83, or Barnhill and Farin, 1981). On the other hand, the interpolant proposed by Schmidt is C^1 and can be represented by a single cubic polynomial on each triangle.

Clearly, a necessary condition for interpolability is that the dimension of the interpolation space equals or exceeds the number of interpolation conditions. If we wish to interpolate to function (but not derivative values) values at the vertices of a triangulation or tetrahedralization the dimension of the piecewise polynomial space must equal or exceed the number of vertices. The following examples establish upper bounds on the degree of the piecewise polynomials potentially useable for interpolation.

Example 12 (Bivariate C^1) (cf. Remark 2 after Theorem 2). Consider the example given by Morgan and Scott, 1975b. In the absence of geometric degeneracies, $\dim(P_2^{1,2}) = 6$. Thus all C^1 piecewise quadratic functions on this triangulation are in fact quadratic. By repeating this triangulation on the subtriangles, triangulations can be constructed with arbitrarily many vertices, for which still $\dim(P_2^{1,2}) = 6$. Thus in general C^1 quadratics can not be used for interpolation. However, on special triangulations, C^1 quadratics may be sufficient. See, for example, Powell, 1973. The situation changes if C^1 cubics are considered. It follows from Table 2 that

$$\dim(S_2^{1,3}) \geq 3B + 2I + 1 > B + I.$$

Thus C^1 cubics provide a sufficient number of parameters for interpolation.

Obviously, more than a sufficient number of parameters is required for interpolability. In principle the smoothness and interpolation conditions may be inconsistent. However, no such case is known. Even for the simplest case of bivariate C^1 piecewise cubic functions it is an open and apparently very difficult question if the Lagrange interpolation problem can always be solved. This problem becomes even more difficult in the following examples.

Example 13 (Bivariate C^2). Similarly as in Example 12, it can be shown that C^2 quartics cannot be used for interpolation. On the Morgan Scott triangulation, $\dim(S_2^{2,4}) = 15$ in the absence of geometric degeneracies. Thus triangulations with arbitrarily many points can be constructed for which all C^2 quartics are in fact quartic. On the other hand, it follows from Table 3 that C^2 quintics provide a sufficient number of parameters for interpolation.

Example 14 (Trivariate C^1). It was shown in Example 5 that on the simple Clough-Tocher split the only C^1 piecewise cubic functions are global cubic polynomials (which have dimension 20). Therefore, the interpolation problem can not be solved in general with C^1 cubics. It follows from Example 3 and Table 6 that tetrahedralizations can be built for which the lower bound on $\dim(S_3^{1,4})$ is smaller than the number of vertices. No case is known, however, where the true dimension of $S_3^{1,4}$ is indeed less than the number of vertices. It is thus an open question if $\dim(S_3^{1,4})$ is in general sufficiently large for interpolation. On the other hand, it follows from

Table 6 that piecewise quintics do always possess sufficiently many parameters for interpolation.

Example 15 (Trivariate C^2). Very little is known in this case. Table 7 shows that $S_3^{2,9}$ is of a sufficiently high dimension for the Lagrange interpolation problem.

ACKNOWLEDGEMENTS

This research was supported by a faculty grant and a sabbatical leave from the University of Utah, by the Department of Energy under Contract No. DEAC02-82-ER-12046 to the University of Utah, by the United States Army under contract No. DAAG29-80-C-0041, and by the University of Dundee. The author has benefited greatly from the stimulating environments provided by the Mathematics Computer Aided Geometric Design Group at the University of Utah, and by the Mathematics Research Center at the University of Wisconsin-Madison.

REFERENCES

P. Alfeld, 1984: *A Trivariate Clough-Tocher Scheme for Tetrahedral Data;* Computer Aided Geometric Design J., North Holland, 1, 169-181.

P. Alfeld, 1985: *A Bivariate C^2 Clough-Tocher Scheme*; Computer Aided Geometric Design J., North Holland, to appear.

R.E. Barnhill, 1985; *Surfaces in Computer Aided Geometric Design*; Computer Aided Geometric Design J., North Holland, to appear.

R.E. Barnhill and G. Farin, 1981: *C^1 Quintic Interpolation over Triangles: Two Explicit Representations*; Int. J. Num. Meth. Engin. 17, 1763-1778.

P.J. Davis, 1975: *Interpolation & Approximation*; Dover Publication.

D.J. Ewing, A.J. Fawkes and J.R. Griffiths, 1970: *Rules governing the numbers of nodes and Elements in a Finite Element Mesh*; Int. J. for Num. Meth. in Eng., 2, 597-601.

G. Farin, 1980: *Bézier Polynomials over Triangles and the Construction of Piecewise C^r Polynomials*; Report TR/91, Dept. of Math., Brunel University, Uxbridge, UK.

R. Franke, 1982: *Scattered Data Interpolation: Tests of some methods*; Math. Comp. 38, 181-200.

A.C. Hearn, 1983: REDUCE *User's Manual, Version 3.0*; The Rand Corporation, Santa Monica, CA 90406.

J. Morgan and R. Scott, 1975a: *A Nodal Basis for C^1 Piecewise Polynomials of Degree* n $\geq$ 5; Math. Comp. 29, 736-740.

J. Morgan and R. Scott, 1975b: *The Dimension of the Space of C^1 Piecewise Polynomials*; Manuscript.

M.J.D. Powell, 1973: *Piecewise quadratic surface fitting for contour plotting*; Proc. Conf. Software for Numerical Mathematics, U. of Technology, Loughborough.

R. Schmidt, 1982: *Eine Methode zur Konstruktion von C^1-Flächen zur Interpolation unregelmässig verteilter Daten*; in: W. Schempp and K. Zeller (ed.), Multivariate Approximation II, 1982, Birkhäuser Verlag, 343-362.

L.L. Schumaker, 1979: *On the dimension of piecewise polynomials in two variables*; in: W. Schempp and K. Zeller (ed.), Multivariate Approximation, 1979, Birkhäuser Verlag, 251-264.

L.L. Schumaker, 1984: *Bounds on the dimension of spaces of multivariate piecewise polynomials*; Rocky Mountain Journal of Mathematics, 14, 251-264.

G. Strang, 1973: *Piecewise polynomials and the finite element method*; Bull. AMS, 79, 1128-1137.

G. Strang, 1974: *The dimension of piecewise polynomials, and one-sided approximation*; Proc. Conf. Numerical Solution of Differential Equations, Dundee (1973), Lecture Notes in Mathematics # 365, Springer Verlag, 144-152.

G. Strang and G.J. Fix, 1973: *An Analysis of the Finite Element Method*; Prentice Hall.

A. Ženišek, 1970: *Interpolation polynomials on the triangle*; Numer. Math. 15, 283-296.

A. Ženišek, 1973: *Polynomial Approximation on Tetrahedrons in the Finite Element Method*; J. Approx. Theory, 7, 334-351.

Peter Alfeld
Department of Mathematics
University of Utah
Salt Lake City
Utah 84112.

F BREZZI

Theoretical and numerical problems in semiconductor devices

1. INTRODUCTION

The aim of this paper is to present some problems arising in the study of field-effect semiconductor devices. The basic equations of the phenomenon are well known and have essentially been accepted by everybody for several years [7]. Nevertheless, the big variety of different applications (value of the coefficients, type and value of the boundary conditions) gives rise to an almost equally big variety of qualitative behaviours, so that a general philosophy is not yet achieved at present. Here I will concentrate on the case of a reverse-biased device. For this case, a slightly "non standard" point of view was introduced about one year ago in [1]. In the present paper the basic ideas of [1] will be presented, together with some more recent results in that direction. A few of them will be presented in forth-coming papers, by other authors, which are still in preparation: they will be referred here as "personal communications".

Important results and basic references in this area (although in other directions) can be found for instance in [3], [4], [5], [6].

An outline of the paper is as follows. In the next section we shall recall the basic equations of [7] and then introduce the reverse-biased case, the simplified model that we are going to analyze, and the basic point of view of [1]. In the third section some of the results obtained so far will be presented. A few numerical results will be shown in the fourth section.

The author is very much indebted with many colleagues (and friends) for several helpful discussions, ideas, and "personal communications": in particular A. Capelo, L. Gastaldi, P.L. Lions, L.D. Marini, N. Nassif and A. Savini.

2. THE MODEL PROBLEMS

Let Ω be a bounded smooth domain in R^2. The basic equations that we are going to study are the following ones: find $\psi(x)$, $p(x)$ and $n(x)$ defined on Ω such that:

$$\operatorname{div}(\varepsilon \operatorname{grad}\psi) = -q(-n + p + C) \text{ in } \Omega,$$

$$\frac{\partial n}{\partial t} - \operatorname{div}(D_n \operatorname{grad} n - \mu_n n \operatorname{grad}\psi) = R \text{ in } \Omega,$$

$$\frac{\partial p}{\partial t} - \operatorname{div}(D_p \operatorname{grad} p + \mu_p p \operatorname{grad}\psi) = R \text{ in } \Omega,$$

+ initial and boundary conditions.

Here $\varepsilon(x)$ is the permittivity of the material, q is the charge of the electron, $C(x)$ is the doping profile. All these are supposed to be known quantities. On the other hand D_n, D_p (diffusion coefficients for electrons and holes, respectively) μ_n, μ_p (mobility coefficients) and R (recombination term) might depend (in general in a nonlinear way) on the unknowns $p(x)$, $n(x)$ and $\operatorname{grad}\psi(x)$. Different models for this dependence are met in the engineering literature. Here we shall start with some drastic simplifications. Namely we shall:

1) study the stationary problem $(\frac{\partial n}{\partial t} = \frac{\partial p}{\partial t} = 0)$,

2) neglect the recombination term $(R = 0)$,

3) assume constant permittivity, diffusion and mobility coefficients (ε, D_n, D_p, μ_n, μ_p constants),

4) assume a piecewise constant doping profile ($\Omega = \Omega_a \cup \Omega_d$ and $C(x) = N_d\, \chi_d(x) - N_a\, \chi_a(x)$ with N_d and N_a constants and $\chi_d(x)$, $\chi_a(x)$ characteristic functions of Ω_d and Ω_a respectively).

This already simplifies the problem quite a bit, but many difficulties are still there. In order to make our device a "reverse-biased" one, we have to say something on the boundary conditions. We assume that we are given on $\partial\Omega \cap \partial\Omega_a$ and on $\partial\Omega \cap \partial\Omega_d$ a finite number of connected components $\Gamma_a^1, \ldots, \Gamma_a^{r_a}, \Gamma_d^1, \ldots, \Gamma_d^{r_d}$. We set:

$$\Gamma_a = \cup\, \Gamma_a^i \qquad ; \quad \Gamma_d = \cup\, \Gamma_d^i \quad ;$$

$$\Gamma_{Dir} = \Gamma_a \cup \Gamma_d \quad ; \quad \Gamma_{Neu} = \partial\Omega - \Gamma_{Dir} \; .$$

We then consider boundary conditions of the following type:

$$n(x) = \bar{n}(x) \quad ; \quad p(x) = \bar{p}(x) \quad ; \quad \psi(x) = \bar{\psi}(x) \quad \text{on } \Gamma_{Dir}$$

$$\frac{\partial n}{\partial \nu} = \frac{\partial p}{\partial \nu} = \frac{\partial \psi}{\partial \nu} = 0 \quad \text{on } \Gamma_{Neu} \quad (\nu = \text{normal})$$

In the applications, in general, $\bar{n}$, $\bar{p}$ and $\bar{\psi}$ will be constant on each connected component of Γ_{Dir}. Moreover the values of $\bar{n}$ and $\bar{p}$ will be strongly related to the values of $\bar{\psi}$ and $C(x)$.

We shall (roughly) say that the device is <u>reverse-biased</u> if:

$$\max_{\Gamma_a} \bar{\psi} < \min_{\Gamma_d} \bar{\psi} .$$

In practical cases, in general, one can observe (if the reverse bias is strong enough) a region surrounding the p-n junction (that is: $\partial\Omega_a \cap \partial\Omega_d$) where both $p(x)$ and $n(x)$ assume a very small value (totally depleted region). See fig. 1.

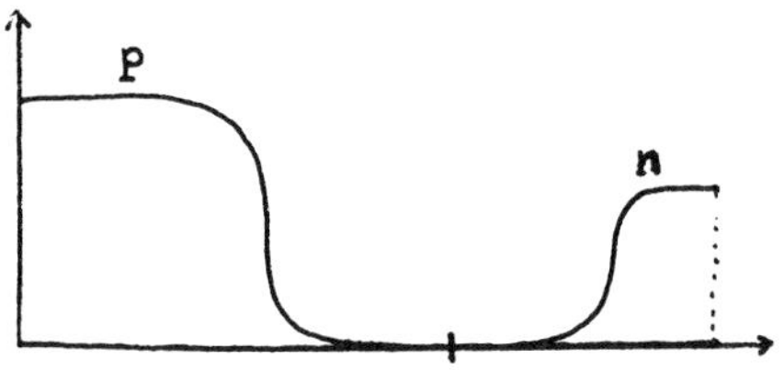

Figure 1

If one of the two sets, say Ω_d, is very thin, one could neglect it, drop the unknown $n(x)$ and consider, on the domain Ω_a only, the simplified problem

$$\begin{aligned} &\operatorname{div}(\varepsilon \operatorname{grad} \psi) = -q(p - N_a) && \text{in } \Omega_a \\ &\operatorname{div}(D_p \operatorname{grad} p + \mu_p\, p \operatorname{grad} \psi) = 0 && \text{in } \Omega_a \\ &p = \bar{p} \text{ on } \Gamma_a \quad ; \quad \psi = \bar{\psi} \text{ on } \Gamma_a \\ &\frac{\partial p}{\partial \nu} = \frac{\partial \psi}{\partial \nu} = 0 \text{ on } \Gamma_{Neu} \cap \partial D_a \end{aligned} \tag{1}$$

One needs boundary conditions on the junction too. We may assume that ψ is computable there

$$\psi = \bar{\psi} \text{ on } \partial\Omega_a \cap \partial\Omega_d \tag{2}$$

For p we may assume

$$\left.\begin{array}{ll} & p = 0 \\ \text{or} & \\ & p = \bar{p} \quad \text{small and known} \end{array}\right\} \quad \text{on } \partial\Omega_a \cap \partial\Omega_d \tag{3}$$

This oversimplified problem, (1) (2) and (3) still retains however considerable difficulties. The reason for that is the big difference in the order of magnitude of the coefficients and, as a consequence, the strong inner layers that $p(x)$ exhibits in practical cases.

In order to analyze this behaviour the following scaling was proposed in [1]

$$\begin{array}{l} \Delta\psi = 1 - p \\ \operatorname{div}(\lambda \operatorname{grad} p + p \operatorname{grad} \psi) = 0 \\ p = 1 \text{ on } \Gamma_a \,,\ \dfrac{\partial p}{\partial \nu} = 0 \text{ on } \Gamma_{Neu} \,,\ p = \beta \text{ on } \Gamma_{ju} \\ \psi = \bar{\psi} \text{ on } \Gamma_a \,,\ \dfrac{\partial \psi}{\partial \nu} = 0 \text{ on } \Gamma_{Neu} \,,\ \psi = \alpha \text{ on } \Gamma_{ju} \end{array} \tag{4}$$

(here $\Gamma_{ju} = \partial\Omega_a \cap \partial\Omega_d$ and we write Γ_{Neu} instead of $\Gamma_{Neu} \cap \partial\Omega_a$).

The constant β is supposed to be bigger than the values of $\bar{\psi}$ on Γ_a (reverse bias). With this scaling $p(x)$ will vary in between 0 and 1.

If we scale the dimension of the domain to be of magnitude 1 the corresponding values for ψ will also, in general, have a reasonable size $(0 \leq \psi \leq 1)$. But a typical λ might be 10^{-9}. The basic point in [1] was to analyze the limit problem for $\lambda = 0$ and, more generally, that all the "a priori estimates" for (4) should be made independent of λ.

We point out that for different problems (forward bias, for instance) a different scaling and a different singular perturbation analysis seem to be more recommended (see [4], [5]).

3. SOME RESULTS

We shall present now some results on problem (4). Some of them have been proved only on the 1-dimensional case, that we rewrite here explicitly:

$$\psi''(x) = 1 - p(x) \text{ in }]0,1[$$

$$(\lambda p' + p\psi')' = 0 \text{ in }]0,1[$$

$$p(0) = 1 \quad p(1) = \beta \qquad 0 \le \beta \le 1$$

$$\psi(0) = 0 \quad \psi(1) = \alpha \qquad \alpha > 0$$

The first theorem is not difficult to prove, via a classical Leray-Shauder fixed point argument (see for instance [6], [3], [5]).

Theorem 1 Assume $\bar{p}$ and $\bar{\psi}$ to be constant (or smooth functions) on each Γ_a^i and on Γ_{ju}, and assume $0 \le \beta \le 1$. Then for any $\lambda > 0$ there exists a solution of (4) (or of (5)). Moreover there exists a constant c independent of λ such that

$$\|\psi\|_{H^1(\Omega_a)} + \|p\|_{L^\infty(\Omega_a)} \le c .$$

In particular $0 \le p \le 1$ in Ω_a.

As far as the smoothness of the solution is concerned, not too much can be expected, for problem (4), due to the mixed Dirichlet-Neumann boundary conditions. However, in the one dimensional case this difficulty is not present.

Theorem 2 The solutions $\psi(x)$ and $p(x)$ of (5) for $\lambda > 0$ are analytic functions. Moreover, there exists a constant c independent of λ such that

$$\|\psi\|_{W^{3,1}} + \|p\|_{W^{1,1}} \le c .$$

The proof is given in [1] in the case $\beta = 0$. The extension to the case $0 < \beta \le 1$ is an exercise.

Other qualitative properties of the solutions can be found for problem (5). For instance one can prove that $p(x)$ is strictly decreasing ([1]) and has at most one inflection point (L. Gastaldi: personal communication).

Let us now consider the limit (if any) of the solutions of problems (4) and (5) as λ goes to zero. For that, from now on, we shall denote by ψ_λ, p_λ the solutions corresponding to a given $\lambda > 0$. From the bounds (6) one gets easily the existence of a subsequence such that

$$\psi_\lambda \rightharpoonup \psi_o \quad \text{in } H^1(\Omega_a), \tag{7}$$

$$p_\lambda \rightharpoonup p_o \quad \text{in } L^\infty(\Omega_a). \tag{8}$$

Moreover, since ψ_λ is actually uniformly in $H^s(\Omega_a)$ for $s < 3/2$, the convergence is strong in (7). One gets also easily that

$$\Delta\psi_o = 1 - p_o \quad \text{in } \Omega_a,$$

$$\operatorname{div}(p_o \nabla\psi_o) = \text{in } \Omega_a,$$

$$\psi_o = \bar{\psi} \text{ on } \Gamma_a \; ; \; \psi_o = \alpha \text{ on } \Gamma_{ju} \; ; \; \frac{\partial\psi_o}{\partial\nu} = 0 \text{ on } \Gamma_{Neu}.$$

It is still not clear what happens of the boundary conditions for p_o. Nevertheless something more can be said on the limit problem. Assume that Γ_a is connected; then it is not restrictive to assume that $\bar{\psi} = 0$ (we have just to shift ψ by a constant). Then one has for $\beta = 0$, that p_o is a characteristic function (i.e. $\int_{\Omega_a} p_o(1-p_o)dx = 0$; P.L. Lions, personal communication). On the other hand, if (for instance) β verifies $\beta = e^{-(\alpha+\lambda^2)/\lambda}$ then one has that ψ_o is the solution of the obstacle problem:

$$\underset{\psi \in K}{\text{minimize}} \int_{\Omega_a} (|\nabla\psi|^2 - 2\psi)dx$$

$$K = \{\phi \mid \phi \in H^1(\Omega), \; \phi|_{\Gamma_a} = 0, \; \phi|_{\Gamma_{ju}} = \alpha, \; \phi \geq 0 \text{ in } \Omega_a\}$$

(see [2]; this in particular implies that $p_o = 1 - \Delta\psi_o$ is a characteristic function).

The situation is much easier for problem (5) (1-dimensional case). In particular one has that, if $\beta = \beta(\lambda) \to 0$ as $\lambda \to 0$, then ψ_o is given, for $0 < \alpha \leq \frac{1}{2}$, by

$$\psi_o(x) = \begin{cases} 0 & \text{for } 0 \leq x \leq \xi \\ (x-\xi)^2/2 & \text{for } \xi \leq x < 1 \end{cases}$$

with $\xi = 1 - \sqrt{2\alpha}$; for $\alpha > \frac{1}{2}$, instead, one has $\psi_o(x) = (x^2/2) + (\alpha - \frac{1}{2})x$. The corresponding behaviour for p_o is deduced from the equation $p_o = 1 - \psi_o''$. It is again a characteristic function, but it is now easy to see that the boundary condition $p(0) = 1$ is lost, in the limit, for $\alpha \geq \frac{1}{2}$ ($p_o \equiv 0$

in this case!). This was proved in [1] for the case $\beta = 0$. The case $\beta(\lambda) \to 0$ can be proved in an identical manner.

4. NUMERICAL RESULTS

The following pictures from [1] show the behaviour of the solutions $\psi(x)$ and $p(x)$ of (5) for $\alpha = 1/8$, $\beta = 0$ and different values of λ. The discretization was done with centered finite differences for the second derivatives and upwinding the first derivative of $p(x)$ in the second equation. The algorithm for computing the discrete solution was, essentially, as follows: to make one sweep of S.O.R. on the second equation using the previous value for $\psi(x)$ and projecting on the barriers $p \geq 0$ and $p \leq 1$ if necessary: then use the updated $p(x)$ in the first equation that was solved for $\psi(x)$ by a direct method. Then go back to the second equation with the new $\psi(x)$ and update $p(x)$ with a new S.O.R. sweep. The method proved to be reasonably fast.

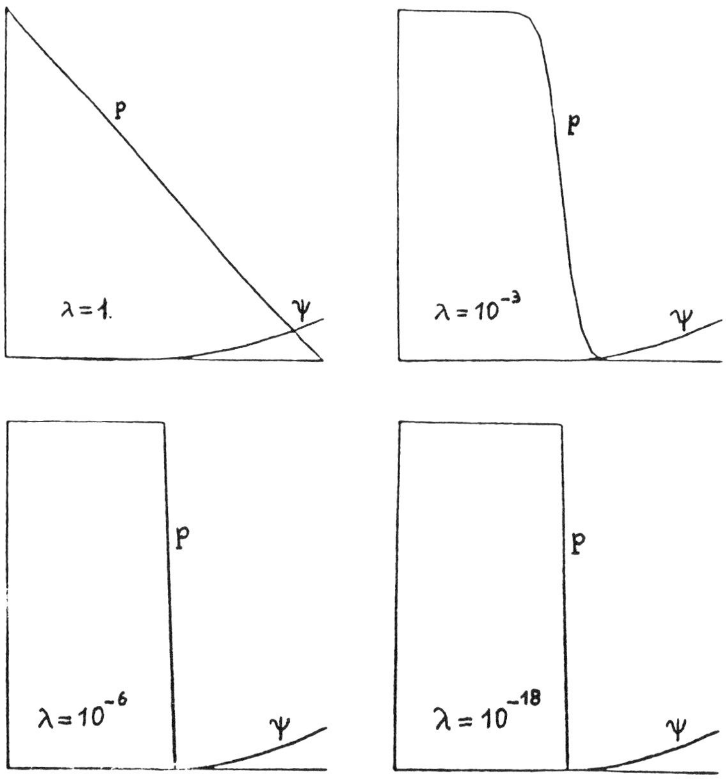

REFERENCES

1. BREZZI, F., CAPELLO, A. and MARINI, L.D. Singular perturbation problems in semiconductor devices, Proceedings of the meeting in Guanajuato, July 1984 - J.P. Hennart ed. (to appear).

2. CAFFARELLI, L. and FRIEDMAN, A. A singular perturbation problem in semiconductors, (to appear).

3. KERKHOVEN, T. On the dependence of the convergence of Gummel's algorithm on the regularity of the solution, Yale Univ. Report DCS/RR366, 1985.

4. MARKOWICH, P.A. A singular perturbation analysis of the fundamental semiconductor equations, SIAM J. Appl. Math., 1984.

5. MARKOWICH, P.A. Book in preparation.

6. MOCK, M. Analysis of Mathematical Models of Semiconductor Devices, Boole, Dublin, 1983.

7. VAN ROOSBROECK, W.V. Theory of flow of electrons and holes in germanium and other semiconductors, Bell Syst. Tech. J. 29 (1950), 560-607.

F. Brezzi
Dipartmento di Meccanica Strutturale dell'
Universita e Istituto di Analisi Numerica del C.N.R.
27100 Pavia
Italy

G J COOPER

An algebraic condition for A-stable Runge-Kutta methods

1. INTRODUCTION

In deciding whether or not a numerical method is suitable for the solution of stiff ordinary differential equations, it is appropriate to determine if it is A-stable. The concept of A-stability, introduced by Dahlquist [5], deals with the behaviour of the method applied to a linear autonomous differential equation. This concept says little about the behaviour of the method applied to a nonlinear system of differential equations and also suffers from the drawback that it can be difficult to determine if a particular method is A-stable.

Other stability definitions have been suggested that partly overcome these defects. The concept of G-stability, due to Dahlquist [6], deals with the stability of linear multistep methods applied to a nonlinear test problem. Butcher [2] used the corresponding autonomous problem to study the stability of Runge-Kutta methods and called this concept B-stability. Burrage and Butcher [1] extended this treatment to include non-autonomous problems and obtained algebraic conditions which guarantee B-stability. The same conditions were obtained by Crouzeix [4]. A method which satisfies these conditions is said to be algebraically stable.

In examining implications of algebraic stability for Runge-Kutta methods, it is necessary to determine whether or not a method can be reduced to another method with fewer stages. Hundsdorfer and Spijker [8] gave a general definition of reducibility and showed that an irreducible Runge-Kutta method is B-stable only if it is algebraically stable. Corresponding necessary and sufficient conditions for A-stability do not seem to have been obtained and, perhaps for this reason, the relationship between A-stability and B-stability has been obscure.

In this article, the usual criterion for A-stability is discussed and a more general condition is obtained. This condition has some defects which arise because the concept of A-stability requires both that there exist a continuous solution of the Runge-Kutta equations and that the modulus of a rational function, given by the method, is bounded by one in the left half plane. A discussion of these requirements leads to sufficient algebraic conditions which happen to be a natural generalization of the algebraic conditions for B-stability. This suggests that there are other stability concepts intermediate between A-stability and B-stability. A stability concept of this type has been described by Cooper [3]. It is noted that the conditions obtained here apply to both reducible and irreducible methods.

Consider an initial value problem for a system of autonomous differential equations

$$x'(t) = f(x(t)), \qquad x(t_0) = \mu,$$

and suppose that an approximation y to $x(t_0+h)$ is required. A Runge-Kutta method for the numerical solution of this problem may be written in the form

$$\begin{aligned} y_i &= \mu + h \sum_{j=1}^{s} a_{ij} f(y_j), \qquad i = 1,2,\dots,s, \\ y &= \mu + h \sum_{i=1}^{s} b_i f(y_i). \end{aligned} \tag{1.1}$$

The method is described concisely by the array

$$\begin{array}{c|c} c & A \\ \hline & b^T \end{array}$$

where A is the real $s \times s$ matrix $[a_{ij}]$ and b^T the real row vector $[b_i]$. The vector $c^T = [c_i]$, defined by $c = Ae$ where $e^T = [1,1,\dots,1]$, is needed to apply the method to non-autonomous systems and plays little role here. In this article, the statement that there exists a solution y of (1.1) implies that there exist $y_1,y_2,\dots,y_s$ which satisfy

$$y_i = \mu + h \sum_{j=1}^{s} a_{ij} (y_j), \qquad i = 1,2,\dots,s. \tag{1.2}$$

This avoids difficulties that may arise when some elements of b are zero.

The concept of A-stability deals with a scalar differential equation

$$x'(t) = \lambda x(t), \qquad \lambda \in \bar{\mathbb{C}}_- = \{\xi : \operatorname{Re} \xi \le 0\}. \tag{1.3}$$

Dahlquist [5] defined A-stability for linear multistep methods and pointed out that the definition could be adapted to examine the behaviour of Runge-Kutta methods. A typical definition follows.

DEFINITION 1.1 Consider a Runge-Kutta method applied to a differential equation (1.3). For given $h > 0$ and initial value μ, suppose there exists a solution y of (1.1). The method is A-stable if for any such problem $|y| \le |\mu|$.

In this definition a strict inequality may be used if attention is restricted to the open set $\mathbb{C}_- = \{\xi : \text{Re } \xi < 0\}$, though this introduces minor technical problems. Note that, if there exist solutions y and w corresponding to initial values μ and ν, then A-stability implies that $|y-w| \le |\mu-\nu|$. In the following there is no loss of generality if $\mu = 1$ is chosen. With this initial value (1.2) gives the linear system

$$(I-zA)Y = e, \qquad z = h\lambda, \tag{1.4}$$

where $Y^T = [y_1, y_2, \ldots, y_s]$. This yields the usual criterion that the method is A-stable if

$$\det(I-zA) \neq 0 \quad \text{and} \quad |1+zb^T(I-zA)^{-1}e| \le 1 \qquad \forall z \in \bar{\mathbb{C}}_-. \tag{1.5}$$

For a particular method, it can be difficult to decide if this criterion is satisfied.

Of more immediate interest is the observation that (1.4) may have a solution Y when $(I-zA)$ is singular and that $y_i : \bar{\mathbb{C}}_- \to \mathbb{C}$, $i = 1,2,\ldots,s$, regarded as functions of z, may be continuous in $\bar{\mathbb{C}}_-$ (and hence analytic in $\mathbb{C}_-$). This suggests that the definition and the criterion (1.5) need modification. To illustrate this point consider the method given by

$$\begin{array}{c|cc} \frac{1}{2} & -\frac{1}{2} & 1 \\ \frac{1}{2} & 0 & \frac{1}{2} \\ \hline & \beta & 1-\beta \end{array}$$

which is of order two and, indeed, reduces to the implicit midpoint rule. For this method $I-zA$ is singular when $z = -2$ and yet (1.4) and (1.1) give

$$Y = \frac{2}{2-z}e \quad \text{and} \quad y(z) = \frac{2+z}{2-z} \qquad \forall z \in \bar{\mathbb{C}}_-,$$

for any β. For this method y is the unique continuous function satisfying (1.1) for each $z \in \bar{\mathbb{C}}_-$. In this sense the method is A-stable although, when $z = -2$, there exists for each $k > 0$ another solution of (1.1) such that $|y| = k$.

The concept of B-stability deals with a system of differential equations

$$x'(t) = f(x(t)), \qquad f : \mathbb{R}^n \to \mathbb{R}^n, \tag{1.6}$$

where, for a given inner product on $\mathbb{R}^n$,

$$\langle f(u)-f(v), u-v \rangle \le 0 \qquad \forall u,v \in \mathbb{R}^n. \tag{1.7}$$

Associated with this inner product is a real positive definite matrix S such that $\langle u,v \rangle = v^T S u$ for all u and v in $\mathbb{R}^n$. The corresponding norm is $\| u \| = \langle u,u \rangle^{\frac{1}{2}}$.

Burrage and Butcher [1] introduced the concept of algebraic stability for Runge-Kutta methods. This concept deals with non-negative definite forms and in this article $M \ge 0$ denotes that the real matrix M is non-negative definite and $M > 0$ denotes that M is positive definite. In the following B denotes the diagonal matrix such that $Be = b$. Hence $B \ge 0$ if and only if each element of b is non-negative. Burrage and Butcher [1] defined a method to be algebraically stable if

$$B \ge 0 \quad \text{and} \quad BA + A^T B - bb^T \ge 0. \tag{1.8}$$

Burrage and Butcher [1] and Crouzeix [4] proved that algebraic stability implies B-stability. Consider a system (1.6) satisfying (1.7) and, for given $h > 0$ and initial values μ and ν, suppose there exist solutions y and w of (1.1). The method is B-stable if for any such problem $\| y-w \| \le \| \mu-\nu \|$. Hundsdorfer and Spijker [8] proved that if an irreducible Runge-Kutta method is B-stable then it is algebraically stable. Applying the usual isomorphism between $\mathbb{R}^2$ and $\mathbb{C}$ it can be shown that B-stability implies A-stability.

In this article conditions for A-stability, which have some similarity to (1.8), are obtained. This similarity is strengthened when some methods, which seem to lack interest, are excluded.

2. THE SCALAR PROBLEM

In this section the definition of A-stability is modified by requiring that the solution of the Runge-Kutta equations be a continuous mapping $y : \bar{\mathbb{C}}_- \to \mathbb{C}$. This gives a weaker criterion than (1.5) and yields an algebraic condition which goes some way to providing a link with algebraic stability. In the remainder of this article the term A-stability refers to the following definition.

DEFINITION 2.1 Consider a Runge-Kutta method applied to a differential equation (1.3) with initial value μ. Suppose there exist continuous $y_i : \bar{\mathbb{C}}_- \to \mathbb{C}$, $i = 1,2,\ldots,s$, satisfying (1.2) for each $z \in \bar{\mathbb{C}}_-$, $z = h\lambda$. The method is A-stable if $|y(z)| \le |\mu|$ for all $z \in \bar{\mathbb{C}}_-$.

To apply this definition, a condition is required that is necessary and sufficient for the existence of a continuous $Y : \bar{\mathbb{C}}_- \to \mathbb{C}^s$ such that

$$(I-zA)Y(z) = e \qquad \forall z \in \bar{\mathbb{C}}_-. \tag{2.1}$$

To this end, consider the sequence $e, Ae, A^2e, \ldots,$ and let A^re be the first vector which is a linear combination of the preceding vectors. Hence

$$o = A^re + p_{r-1}A^{r-1}e + \ldots + p_1Ae + p_0e, \qquad 1 \le r \le s, \tag{2.2}$$

and the $s \times r$ matrix $M = [e, Ae, \ldots, A^{r-1}e]$ is of column rank r. The minimal polynomial p_e of A for e, which divides the minimal polynomial of A, is the monic polynomial defined by $p_e(z) = z^r + p_{r-1}z^{r-1} + \ldots + p_1z + p_0$. Here, it is more appropriate to use the transformed polynomial p, given by

$$p(z) = 1 + p_{r-1}z + \ldots + p_1z^{r-1} + p_0z^r, \tag{2.3}$$

and the associated non-singular $r \times r$ matrix P,

$$P = \begin{bmatrix} 1 & p_{r-1} & p_{r-2} & \cdots & p_2 & p_1 \\ 0 & 1 & p_{r-1} & & p_3 & p_2 \\ 0 & 0 & 1 & & p_4 & p_3 \\ \vdots & & & & & \vdots \\ 0 & 0 & 0 & & 1 & p_{r-1} \\ 0 & 0 & 0 & \cdots & 0 & 1 \end{bmatrix}.$$

In the proof of the following theorem it is shown that there is a unique continuous Y which satisfies (2.1) if and only if p is not zero in $\bar{\mathbb{C}}_-$.

THEOREM 2.1 Let $Z : \mathbb{C} \to \mathbb{C}^r$ be the column vector defined by $Z^T(z) = [1, z, \ldots, z^{r-1}]$. A Runge-Kutta method is A-stable if and only if

$$p(z) \ne 0 \quad \text{and} \quad \left|1 + \frac{z}{p(z)} b^TMPZ(z)\right| \le 1 \qquad \forall z \in \bar{\mathbb{C}}_-.$$

<u>Proof</u>. Since $M = [e, Ae, \ldots, A^{r-1}e]$ the linear dependence (2.2) and some manipulation give

$$zAMPZ(z) = MPZ(z) - p(z)e \qquad \forall z \in \mathbb{C}. \tag{2.4}$$

It follows that a solution of (2.1) is given by

$$Y(z) = \frac{1}{p(z)} MPZ(z), \qquad p(z) \neq 0. \tag{2.5}$$

If $\det(I-zA) \neq 0$ this solution is unique. Suppose that $\det(I-zA) = 0$ but $p(z) \neq 0$. At such a point, other solutions exist but Y defined by (2.5) is the unique continuous solution. If $p(z) = 0$ a solution may not exist. If there is a solution then Y is continuous only if $MPZ(z) = 0$ and this implies that the column rank of M is less than r. Hence there is a continuous $Y : \bar{\mathbb{C}}_- \to \mathbb{C}^r$ satisfying (2.1) if and only if p is not zero in $\bar{\mathbb{C}}_-$. This mapping is unique and is given by (2.5). The theorem follows from definition (2.1) since $y(z) = 1 + zb^TY(z)$ when $\mu = 1$.

It has been remarked that a strict inequality could be used in the definition of A-stability. The next result amplifies this comment.

COROLLARY For a consistent A-stable method, $|y(z)| < |\mu|$ for all $z \in \mathbb{C}_-$ $(\mu \neq 0)$.

Proof. Let $\mu = 1$ and suppose that $|y(z)| = 1$ for some $z \in \mathbb{C}_-$. Since y is analytic and $|y(z)| \leq 1$ in $\mathbb{C}_-$, the maximum modulus principle gives $|y(z)| = 1$ for all $z \in \mathbb{C}_-$. Hence $b^TMPZ(z) = 0$ for all $z \in \mathbb{C}_-$. Since P is non-singular this gives $b^TM = 0$ and therefore $b^Te = 0$ which contradicts the consistency condition $b^Te = 1$.

Let $\bar{p}$ be the polynomial defined by $\bar{p}(z) = p(-z)$. Since p is not zero in $\bar{\mathbb{C}}_-$ if and only if all zeros of $\bar{p}$ lie in $\mathbb{C}_-$, the Routh-Hurwitz criterion gives a necessary and sufficient algebraic condition for p to be non-zero in $\bar{\mathbb{C}}_-$. In the next result the inequality in theorem (2.1) is replaced by an algebraic condition.

THEOREM 2.2 A method is A-stable if and only if p is not zero in $\bar{\mathbb{C}}_-$ and there exists a real symmetric D such that $M^T(De-b) = 0$ and

$$Z^T(\bar{z})P^TM^T(DA + A^TD - bb^T)MPZ(z) \geq 0 \qquad \forall z = ix, \quad x \text{ real}.$$

This inequality is independent of D.

Proof. Suppose that p is not zero in $\bar{\mathbb{C}}_-$. Since y is analytic in $\mathbb{C}_-$

and continuous in $\bar{\mathbb{C}}_-$, it follows from the maximum modulus principle that the method is A-stable if and only if $|y(z)| \leq 1$ for all $z = ix$. This is equivalent to the condition $\phi(z) \geq 0$ for all $z = ix$, where

$$\begin{aligned}\phi(z) &= |p(z)|^2 - |p(z)y(z)|^2 \\ &= |p(z)|^2 - [p(\bar{z}) + \bar{z}Z^T(\bar{z})P^TM^Tb][p(z) + zb^TMPZ(z)] \\ &= -|z|^2Z^T(\bar{z})P^TM^Tbb^TMPZ(z) - zp(\bar{z})b^TMPZ(z) - \bar{z}p(z)Z^T(\bar{z})P^TM^Tb. \quad (2.6)\end{aligned}$$

Let D be any real symmetric $s \times s$ matrix such that $M^T(De-b) = 0$. Since $z + \bar{z} = 0$ for all $z = ix$, equations (2.4) and (2.6) give

$$\phi(z) = |z|^2Z^T(\bar{z})P^TM^T(DA + A^TD - bb^T)MPZ(z) \qquad \forall z = ix.$$

To illustrate this result consider, again, the method given by

$\frac{1}{2}$	$-\frac{1}{2}$	1
$\frac{1}{2}$	0	$\frac{1}{2}$
	β	$1-\beta$

.

In this case $M = e$ and $p(z) = 1 - \frac{1}{2}z$ which is not zero for $z \in \bar{\mathbb{C}}_-$. The choice $D = B$ gives $M^T(DA + A^TD - bb^T)M = 0$ so that the conditions given in theorem 2.2 hold. Next consider the trapezoidal rule

0	0	0
1	$\frac{1}{2}$	$\frac{1}{2}$
	$\frac{1}{2}$	$\frac{1}{2}$

, $M = \begin{bmatrix} 1 & 0 \\ 1 & 1 \end{bmatrix}$,

where, again, $p(z) = 1 - \frac{1}{2}z$. The choice $D = B$ gives

$$Z^T(-ix)P^TM^T(DA + A^TD - bb^T)MPZ(ix) = [1, -ix] \begin{bmatrix} 0 & \frac{1}{4} \\ \frac{1}{4} & 0 \end{bmatrix} \begin{bmatrix} 1 \\ ix \end{bmatrix} = 0$$

which shows that the method is A-stable. However, the choice $D = bb^T$ gives a more direct confirmation of A-stability since, in this case, $DA + A^TD - bb^T = 0$.

A more elaborate example is the method described by

$$\begin{array}{c|ccc}
\frac{3-\sqrt{3}}{6} & \frac{1}{4} & \alpha(1+u) & \alpha(1+u) \\
\frac{3+\sqrt{3}}{6} & 0 & \frac{3}{8}-\alpha(1-u) & \frac{3}{8}-\alpha(1+u) \\
\frac{3+\sqrt{3}}{6} & 4\beta & \frac{\alpha-\beta}{2}+\alpha u & \frac{\alpha-\beta}{2}-\alpha u \\
\hline
 & \frac{1}{2} & \frac{1}{4} & \frac{1}{4}
\end{array}, \qquad \alpha = \frac{3-2\sqrt{3}}{24}, \quad \beta = \frac{3+2\sqrt{3}}{24}.$$

This method, given by Burrage and Butcher [1], is of order four and is A-stable for all $u \geq 0$. In this case $p(z) = 1 - \frac{1}{2}z + \frac{1}{24}(2+u)z^2 - \frac{1}{48}uz^3$ and, applying the Routh-Hurwitz criterion, p is not zero in $\bar{\mathbb{C}}_-$ if and only if $u \geq 0$. It can be shown that

$$D = \begin{bmatrix} \frac{1}{2} & 0 & 0 \\ 0 & \frac{1}{4}-d & d \\ 0 & d & \frac{1}{4}-d \end{bmatrix}, \qquad d = \frac{1-(7-4\sqrt{3})u}{8},$$

gives $DA + A^TD - bb^T = 0$. Note that $D \geq 0$ for all $u \geq 0$.

These examples suggest that if a method is A-stable then it may be possible always to choose D symmetric so that $M^T(De-b) = 0$ and $M^T(DA + A^TD - bb^T)M \geq 0$. Such a result would provide a direct link between A-stability and B-stability. To examine this possibility it is shown next that D can be chosen so that $P^TM^T(DA + A^TD - bb^T)MP$ is diagonal.

THEOREM 2.3 For any Runge-Kutta method there is a symmetric matrix D such that $M^T(De-b) = 0$ and $P^TM^T(DA + A^TD - bb^T)MP$ is diagonal. This diagonal matrix Q is unique. Let $q_1, q_2, \ldots, q_r$ be the diagonal elements. The method is A-stable if and only if p is not zero in $\bar{\mathbb{C}}_-$ and

$$q(x) = q_1 + q_2x + \ldots + q_rx^{r-1} \geq 0 \qquad \forall x > 0.$$

Proof. Let $Q(D) = P^TM^T(DA + A^TD - bb^T)MP$ and let R be the $r \times r$ matrix

$$R = \begin{bmatrix} -p_{r-1} & -p_{r-2} & \cdots & -p_1 & -p_0 \\ 1 & 0 & & 0 & 0 \\ 0 & 1 & & 0 & 0 \\ \vdots & & & & \\ 0 & 0 & & 1 & 0 \end{bmatrix}. \qquad (2.7)$$

The identity $AMP = MPR$ gives

$$Q(D) = WR + R^TW - aa^T, \qquad W = P^TM^TDMP, \qquad a = P^TM^Tb. \tag{2.8}$$

Let $e_1, e_2, \ldots, e_r$ be the natural basis for $\mathbb{C}^r$. Since $MPe_1 = e$ it follows that $M^T(De-b) = 0$ if and only if $e_1^TW = a^T = [a_1, a_2, \ldots, a_r]$. The remaining elements of the symmetric matrix $W = [w_{ij}]$ have to be chosen so that $Q(D)$ is diagonal. This gives

$$w_{i+1j} + w_{ij+1} = a_ip_{r-j} + a_jp_{r-i} + a_ia_j, \qquad \begin{array}{l} j = i+1, i+2, \ldots, r, \\ i = 1, 2, \ldots, r-1, \end{array} \tag{2.9}$$

where $w_{ir+1} = 0$ and $w_{1i} = a_i$, $i = 1, 2, \ldots, r$. These relations uniquely determine W and hence Q. Now partition M and choose D so that

$$M = \begin{bmatrix} M_1 \\ M_2 \end{bmatrix}, \qquad D = \begin{bmatrix} D_1 & 0 \\ 0 & 0 \end{bmatrix},$$

where M_1 and D_1 are $r \times r$ matrices. Since M_1 and P are non-singular it follows that for any W there exists D_1 such that $W = P^TM^TDMP$. Note that D is symmetric but is unique only if $r = s$. Since theorem 2.2 holds for any symmetric D with $M^T(De-b) = 0$, the method is A-stable if and only if p is not zero in $\bar{\mathbb{C}}_-$ and $q(x) \geq 0$ $\forall x > 0$.

It is not necessary to choose D so that $Q(D)$ is diagonal. If $Q(D) \geq 0$ and p is not zero in $\bar{\mathbb{C}}_-$ then the method is A-stable. Now consider an A-stable method. It follows from theorem 2.3 that the diagonal matrix $Q \geq 0$ if $r \leq 2$, but this is not a necessary condition for A-stability when $r \geq 3$. Nevertheless, it may be possible to choose D so that $Q(D) \geq 0$ and it can be shown that this is so when $r = 3$. This is also true when $r = 4$ if the method is of order at least two. To show that this is not always the case consider the first order method

$$\begin{array}{c|cccc} 1 & 1 & 0 & 0 & 0 \\ 0 & 1 & 1 & -1 & -1 \\ 2 & 0 & 0 & 1 & 1 \\ -1 & -1 & -1 & 0 & 1 \\ \hline & \frac{6}{5} & \frac{4}{5} & -\frac{2}{5} & -\frac{3}{5} \end{array}, \qquad M = \begin{bmatrix} 1 & 1 & 1 & 1 \\ 1 & 0 & 0 & 2 \\ 1 & 2 & 1 & -1 \\ 1 & -1 & -2 & -3 \end{bmatrix},$$

where $p(z) = 1 - 4z + 5z^2 - 3z^3 + z^4$. The Routh-Hurwitz criterion shows that p is not zero in $\bar{\mathbb{C}}_-$ and a short calculation gives $a^T = b^TMP = [1, -3, 3, -1]$.

The diagonal elements of Q, obtained from (2.8) and (2.9), give

$$q(x) = 1 - x - x^2 + x^3 = (1-x)^2(1+x) \geq 0 \qquad \forall x > 0,$$

so that the method is A-stable. To show that W cannot be chosen so that $We_1 = a$ and $Q(D) \geq 0$, let $We_1 = a$. Then (2.8) gives

$$Q(D) = \begin{bmatrix} 1 & w_{22}-11 & w_{23}+11 & w_{24}-4 \\ w_{22}-11 & 2w_{23}+21 & w_{33}+w_{24}-15 & w_{34}+5 \\ w_{23}+11 & w_{33}+w_{24}-15 & 2w_{34}+9 & w_{44}-3 \\ w_{24}-4 & w_{34}+5 & w_{44}-3 & 1 \end{bmatrix}.$$

Let $x^T = [0,-1,0,\alpha]$ and $y^T = [\alpha,0,-1,0]$. If $Q(D) \geq 0$ then

$$x^TQ(D)x = \alpha^2 - 10\alpha + 21 + 2w_{23} - 2\alpha w_{34} \geq 0,$$

$$y^TQ(D)y = \alpha^2 - 22\alpha + 9 - 2\alpha w_{23} + 2w_{34} \geq 0,$$

and it can be shown that w_{23} and w_{34} cannot be chosen so that both these inequalities hold for all real α.

The next result shows that if it is possible to choose D so that $Q(D) \geq 0$ then D may be chosen to be non-negative definite.

THEOREM 2.4 Suppose that p is not zero in $\bar{\mathbb{C}}_-$ and that there is a symmetric D such that $M^T(DA + A^TD)M \geq 0$. Then $M^TDM \geq 0$.

Proof. Let $W = P^TM^TDMP$. Then $WR + R^TW \geq 0$ and it has to be shown that $W \geq 0$. Let $WR + R^TW = U$ and note that the eigenvalues of $-R$ lie in $\mathbb{C}_-$. The unique solution of this equation, given by Hahn [7], is

$$W = \int_0^\infty e^{-tR^T}\, U\, e^{-tR}\, dt.$$

Since $U \geq 0$ this gives $W \geq 0$.

3. LINEAR AUTONOMOUS SYSTEMS

This section deals with the behaviour of A-stable Runge-Kutta methods applied to linear autonomous systems of differential equations. Let $x : \mathbb{R} \to \mathbb{R}^n$ be the solution of an initial value problem

$$x'(t) = J\,x(t), \qquad x(t_0) = \mu, \qquad x^T = [x_1,x_2,\ldots,x_n], \tag{3.1}$$

and suppose that, for each initial value μ, there exists a K such that

$|x_i(t)| \le K$ for $i = 1,2,\ldots,n$, for all $t \ge t_0$. An equivalent statement is that there is a T such that each $|x_i(t)|$, $i = 1,2,\ldots,n$, is non-increasing for all $t \ge T$. A necessary and sufficient condition is that each eigenvalue of the real $n \times n$ matrix J lies in $\bar{\mathbb{C}}_-$ and, if λ is an eigenvalue on the boundary of $\bar{\mathbb{C}}_-$, of multiplicity m, there are m linearly independent eigenvectors belonging to λ. This set of matrices is denoted by $\mathbb{L}^n_-$. It is known that $J \in \mathbb{L}^n_-$ if and only if there exists a real symmetric matrix S such that

$$S > 0 \quad \text{and} \quad SJ + J^T S \le 0. \tag{3.2}$$

A norm on $\mathbb{R}^n$ is defined by $\| u \|_S = (u^T S u)^{\frac{1}{2}}$ for all $u \in \mathbb{R}^n$. It can be shown that if x is the solution of an initial value problem (3.1), for which (3.2) holds, then $\| x(t) \|_S$ is non-increasing for all $t \ge t_0$. More details are given by Hahn [7].

To examine the behaviour of a Runge-Kutta method, applied to a system of differential equations, it is convenient to use a tensor product notation. Let $M = [m_{ij}]$ be any $p \times q$ matrix and let R be any $r \times s$ matrix. The tensor product $M \otimes R$ is defined by

$$M \otimes R = \begin{bmatrix} m_{11}R & m_{12}R & \cdots & m_{1q}R \\ m_{21}R & m_{22}R & & m_{2q}R \\ \vdots & & & \vdots \\ m_{p1}R & m_{p2}R & \cdots & m_{pq}R \end{bmatrix}.$$

An account of properties of this product is given by Lancaster [9]. In particular, $(M \otimes R)^T = M^T \otimes R^T$ and $(M \otimes R)(N \otimes S) = (MN) \otimes (RS)$ for conformable matrices. To avoid excessive use of parentheses, the tensor product is given priority over matrix multiplication. Since no ambiguities arise, in the following I denotes the identity matrix of dimension specified by the context.

Now consider a Runge-Kutta method applied to the initial value problem (3.1) and suppose, with no loss of generality, that the step length $h = 1$. For this problem equations (1.1) may be written in the form

$$(I - A \otimes J)Y = e \otimes \mu, \qquad Y^T = [y_1^T, y_2^T, \ldots, y_s^T], \tag{3.3}$$

$$y = \mu + (b^T \otimes J)Y. \tag{3.4}$$

Now consider the linear system of equations

$$(I-R\otimes J)X = e_1\otimes\mu, \qquad X^T = [x_1^T, x_2^T, \ldots, x_r^T],$$

where R is the $r \times r$ matrix given by (2.7). Suppose that $J \in \mathbb{L}_-^n$. If p is not zero in $\bar{\mathbb{C}}_-$ the matrix $p(J)$ is non-singular and there is a unique solution given by

$$x_i = p(J)^{-1}J^{i-1}\mu, \qquad i = 1,2,\ldots,r. \tag{3.5}$$

The identity $AMP = MPR$ gives $(MP)\otimes I\ (I-R\otimes J)X = (I-A\otimes J)(MP)\otimes I\ X$ and since $MPe_1 = e$ it follows that $Y = (MP)\otimes I\ X$ satisfies (3.3). Further, this is the unique solution such that each $y_i : \mathbb{L}_-^n \to \mathbb{R}^n$, $i = 1,2,\ldots,s$, is a continuous mapping into $\mathbb{R}^n$ from $\mathbb{L}_-^n$, regarded as a subset of the space of linear mappings of $\mathbb{R}^n$ into $\mathbb{R}^n$. There is no continuous solution if $p(z) = 0$ for some $z \in \bar{\mathbb{C}}_-$. Corresponding to this solution

$$y = \mu + (b^TMP)\otimes J\ X = [I + p(J)^{-1} \sum_{i=1}^{r} a_iJ^i]\mu \tag{3.6}$$

where $b^TMP = a^T = [a_1, a_2, \ldots, a_r]$.

THEOREM 3.1 Consider an A-stable Runge-Kutta method. If $J \in \mathbb{L}_-^n$ then $\| y \|_S \le \| \mu \|_S$. If the method is consistent and all eigenvalues of J lie in $\mathbb{C}_-$ then $\| y \|_S < \| \mu \|_S$ for $\mu \neq 0$.

<u>Proof</u>. Note that (3.6) gives $y = R(J)\mu$ where R is the rational function defined by $R(z) = 1 + p(z)^{-1}za^TZ(z)$. Since (3.2) gives $\langle Ju,u\rangle \le 0$ $\forall u \in \mathbb{R}^n$, the left half plane $\bar{\mathbb{C}}_-$ is a spectral set for J. Since theorem (2.1) gives $|R(z)| \le 1$ $\forall z \in \bar{\mathbb{C}}_-$, it follows from the von Neumann theory of spectral sets [10] that $\| R(J) \| \le 1$. Now suppose that the method is consistent, so that $|R(z)| < 1$ $\forall z \in \mathbb{C}_-$, and that all eigenvalues of J lie in $\mathbb{C}_-$. Let $\alpha < 0$ and consider $\| J-\alpha I \|^2$, the largest eigenvalue of

$$(J-\alpha I)^TS(J-\alpha I) = \alpha^2I - \alpha(SJ + J^TS) + J^TSJ.$$

Since $SJ + J^TS < 0$ there is a real unitary matrix U such that $U^T(SJ + J^TS)U$ is diagonal with negative diagonal elements. Let $\varepsilon < 0$ be the largest of these elements. It follows from the Gerschgorin circle theorem that $\| J - \alpha I \|^2 \le \alpha^2 - \alpha\varepsilon + c$, where $c \ge 0$ does not depend on α. Hence

$$\| J - \alpha I \| \le \beta, \qquad \beta = |\alpha| + \frac{1}{\alpha}, \qquad \alpha = \frac{2 + c}{\varepsilon},$$

and the disc $|z - \alpha| \le \beta$ is a spectral set for J. This disc lies in $\mathbb{C}_-$ and so $|R(z)| \le k < 1$ on this disc. Therefore $\| R(J) \| \le k < 1$.

It is interesting to observe that a slightly less general result can be proved by different means. This gives a direct relation between $\| y \|_S$ and $\| \mu \|_S$.

THEOREM 3.2 Suppose that p is not zero in $\bar{\mathbb{C}}_-$ and there is a symmetric D such that

$$M^T(De - b) = 0 \quad \text{and} \quad M^T(DA + A^TD - bb^T)M \ge 0.$$

If $J \in \mathbb{L}_-^n$ then $\| y \|_S \le \| \mu \|_S$. If $M^TDM > 0$ and all eigenvalues of J lie in $\mathbb{C}_-$ then $\| y \|_S < \| \mu \|_S$ for $\mu \ne 0$.

<u>Proof</u>. Consider any initial value problem (3.1) for which (3.2) holds. Equation (3.3) gives $Y^TD\otimes(J^TS)\,(I{-}A\otimes J)\,Y = Y^T\,(De)\otimes(J^TS)\,\mu$ and hence

$$Y^T\,(DA)\otimes(J^TSJ)\,Y = Y^T\,D\otimes(J^TS)\,Y - Y^T\,(De)\otimes(J^TS)\,\mu.$$

Since $x^TNx = x^TN^Tx$ and since D and S are symmetric, this gives

$$Y^T\,(DA + A^TD)\otimes(J^TSJ)\,Y = Y^T\,D\otimes(SJ + J^TS)\,Y - 2Y^T\,(De)\otimes(J^TS)\,\mu. \tag{3.7}$$

On the other hand, (3.4) gives

$$y^TSy = \mu^TS\mu + Y^T\,(bb^T)\otimes(J^TSJ)\,Y + 2Y^T\,b\otimes(J^TS)\,\mu. \tag{3.8}$$

Let $W = P^TM^TDMP$ and $Q(D) = P^TM^T(DA + A^TD - bb^T)MP$ and note that $Y = (MP)\otimes I\,X$ and $M^T(De - b) = 0$. Addition of (3.7) and (3.8) gives

$$y^TSy = \mu^TS\mu - X^T\,Q(D)\otimes(J^TSJ)\,X + X^T\,W\otimes(SJ + J^TS)\,X. \tag{3.9}$$

Since P is non-singular $Q(D) \ge 0$ and since $J^TSJ \ge 0$ also, $Q(D)\otimes(J^TSJ) \ge 0$. Since $M^T(DA + A^TD)M \ge 0$ and p is not zero in $\bar{\mathbb{C}}_-$ it follows from theorem 2.4 that $W \ge 0$. But $SJ + J^TS \le 0$ so that $W\otimes(SJ + J^TS) \le 0$. Hence (3.9) gives $y^TSy \le \mu^TS\mu$. If all eigenvalues of J

lie in $\mathbb{C}_-$ then $SJ + J^TS < 0$ so that $W \otimes (SJ + J^TS) < 0$ if $M^TDM > 0$.

The conditions given in this theorem hold for many A-stable methods but it is often difficult to check that p is not zero in $\bar{\mathbb{C}}_-$ by using the Routh-Hurwitz criterion. The next result provides an alternative criterion.

THEOREM 3.3 The polynomial p is not zero in $\bar{\mathbb{C}}_-$ if and only if there is a $\bar{b} \in \mathbb{R}^s$, not orthogonal to any eigenvector of A, and a symmetric $\bar{D}$ such that

$$M^T\bar{D}M > 0 \quad \text{and} \quad M^T(\bar{D}A + A^T\bar{D} - \bar{b}\bar{b}^T)M \geq 0.$$

Proof. Note that p is not zero in $\bar{\mathbb{C}}_-$ if and only if all eigenvalues of the $r \times r$ matrix R have positive real parts. Suppose this is so. Then there is a $W > 0$ such that $WR + R^TW > 0$. Hence, given any $a \in \mathbb{R}^r$, there is a $W > 0$ such that $WR + R^TW - aa^T \geq 0$. Since P is non-singular and M is of column rank r there is an $s \times s$ symmetric $\bar{D}$ such that $W = P^TM^T\bar{D}MP$ and a $\bar{b}$ such that $P^TM^T\bar{b} = a$. Thus $M^T\bar{D}M > 0$ and the identity $AMP = MPR$ gives $M^T(\bar{D}A + A^T\bar{D} - \bar{b}\bar{b}^T)M \geq 0$. Now suppose that there is a $\bar{b}$ not orthogonal to any eigenvector of A. That is, $a = P^TM^T\bar{b}$ is not orthogonal to any eigenvector of R. Suppose also, that $WR + R^TW - aa^T \geq 0$ where $W = P^TM^T\bar{D}MP > 0$. Since $WR + R^TW \geq 0$ the eigenvalues of R have non-negative real parts. Let λ be an eigenvalue and x an eigenvector belonging to λ. Then

$$x^H(WR + R^TW - aa^T)x = (\lambda + \bar{\lambda})x^HWx - |a^Tx|^2 \geq 0.$$

But a is not orthogonal to x and $x^HWx > 0$. This gives $\operatorname{Re}\lambda > 0$.

Compare this result with theorem 3.2 and note that theorem 2.4 gives $M^TDM \geq 0$. Often $M^TDM > 0$ and it may suffice to choose $\bar{b} = b$ as well as $\bar{D} = D$.

Acknowledgement. The author thanks W.H. Hundsdorfer who pointed out an error in the original manuscript and drew attention to the theory of spectral sets.

REFERENCES

1. BURRAGE, K. and BUTCHER, J.C. Stability criteria for implicit Runge-Kutta methods, SIAM J. Numer. Anal. 16 (1979) 46-57.
2. BUTCHER, J.C. A stability property of implicit Runge-Kutta methods, BIT 15 (1975) 358-361.
3. COOPER, G.J. A modification of algebraic stability for implicit Runge-Kutta methods, submitted to IMA J. Numer. Anal.
4. CROUZEIX, M. Sur la B-stabilité des méthodes de Runge-Kutta, Numer. Math. 32 (1979) 75-82.
5. DAHLQUIST, G. A special stability problem for linear multistep methods, BIT 3 (1963) 27-43.
6. DAHLQUIST, G. On stability and error analysis for stiff non-linear problems, Report NA 75.08 Dept. of Inf. Proc., Royal Inst. of Tech., Stockholm, 1975.
7. HAHN, W. Stability of Motion, Springer-Verlag, Berlin, 1967.
8. HUNDSDORFER, W.H. and SPIJKER, M.N. A note on B-stability of Runge-Kutta methods, Numer. Math. 36 (1981) 319-331.
9. LANCASTER, P. Theory of Matrices, Academic Press, New York, 1969.
10. RIESZ, F. and SZ.-NAGY, B. Functional Analysis, Ungar Publishing Co., New York, 1955.

G.J. Cooper
School of Mathematical and Physical Sciences
The University of Sussex
Brighton BN1 9QH,
England

S FRIEDLAND, J NOCEDAL & M L OVERTON

Four quadratically convergent methods for solving inverse eigenvalue problems

1. INTRODUCTION

Consider the following inverse eigenvalue problem:

IEP1: Given real symmetric $n \times n$ matrices $A_0, A_1, \ldots, A_n$ and real numbers $\lambda_1^* \leq \lambda_2^* \leq \ldots \leq \lambda_n^*$, find $c \in \mathbb{R}^n$ such that the eigenvalues of

$$A(c) = A_0 + \sum_{k=1}^{n} c_k A_k \tag{1.1}$$

are $\lambda_1^*, \ldots, \lambda_n^*$.

This problem, or variations on it, occurs in many applications. In this paper we describe four numerical methods for solving IEP1, all of which are related to Newton's method and are generally quadratically convergent. One of these methods is new. We begin by deriving the methods in the case where the given eigenvalues are distinct. We then explain how the methods behave when the set $\{\lambda_i^*\}_1^n$ includes multiple eigenvalues, and we show that in this case the problem IEP1 should be reformulated and the methods modified. We have described all these methods in more detail in another paper [9]. We have included there a general convergence analysis which covers both the distinct and multiple eigenvalue cases. No convergence theorems are given in this paper.

We begin by mentioning some applications where Problem IEP1 and its variations arise. One classical inverse problem involves the Sturm-Liouville equation. Given the differential equation

$$- u''(x) + p(x)u(x) = \lambda u(x) ,$$

with Dirichlet boundary conditions, one wishes to determine the potential function $p(x)$ from the spectrum $\{\lambda_i^*\}_1^\infty$. When the equation is discretized, it takes the form IEP1, where A_0 represents the second-order difference operator and

$$A_k = \begin{bmatrix} & & \\ & 1 & \\ & & \end{bmatrix} \begin{matrix} \\ k \\ \\ \end{matrix} \quad . \tag{1.2}$$

$$\qquad\qquad k$$

Hald [13] is an excellent reference for discrete and continuous inverse Sturm-Liouville problems. Downing and Householder [7] give the name <u>additive inverse eigenvalue problem</u> to IEP1 when $\{A_k\}$ have the form (1.2), i.e.

$$A(c) = A_0 + \mathrm{Diag}(c_k) \ .$$

In the discrete inverse Sturm-Liouville problem the given eigenvalues are always distinct. A simple example where the given eigenvalues are multiple is the problem of communality [14]. Here one is given a symmetric matrix A_0, and one wishes to add diagonal elements to A_0 to make the resulting matrix have the lowest possible rank. In other words, one wishes to set as many eigenvalues equal to zero as possible. Since neither the rank nor the nonzero eigenvalues are known, the problem is not in the form IEP1, but one can consider guessing the rank, and hence the multiplicity of the zero eigenvalue. In some cases this is enough to locally determine a solution, as will be shown later.

A closely related problem is studied by Fletcher [8]. Here one wishes to subtract as much as possible from the diagonal of a positive definite matrix A_0, with the constraint that the matrix remains positive semi-definite, i.e. all the eigenvalues remain nonnegative. As in the problem of communality there is usually a multiple zero eigenvalue at the solution.

More generally, consider minimizing some smooth function of c subject to the constraint that all the eigenvalues of A(c) remain between given fixed lower and upper bounds. Problems of this kind arise in structural engineering, where the natural frequencies of a structure are required to be within certain safety limits (see [19]) and in control engineering, where the singular values of a complex transfer matrix G(c) (eigenvalues of $G(c)^H G(c)$) are required to be in a given interval (see [17]). Note that multiple eigenvalues tend to occur at the solution of such problems, since the minimization objective may drive several eigenvalues to the same bound. A related problem has a "min-max" formulation, where the maximum eigenvalue of A(c) is to be minimized. For example, in control engineering one sometimes wishes to minimize the spectral norm, i.e. the largest singular value, of a transfer matrix G(c); see [24]. Again, multiple eigenvalues tend to arise at the solution, although in this case the maximum value is not known in advance but must be determined during the minimization process.

None of these minimization problems has the form IEP1. However, the relationship between numerical methods for constrained minimization and those for solving nonlinear equations is well known. We shall therefore discuss numerical methods only for the equality model problem IEP1 (and the reformulated version for multiple eigenvalues). We think that a good understanding of the model problem is an essential prerequisite for designing methods for the more general minimization problems just mentioned. Furthermore, we think that it is not difficult to see how to extend the ideas presented here to cover such minimization problems. Other generalizations to the model problem may also be appropriate; in particular, A(c) is a nonlinear matrix function in many applications. We shall assume here that A(c) is the affine function (1.1), but it is not difficult to generalize our remarks to the nonlinear case.

In the remainder of the paper we shall discuss the formulation and local analysis of numerical methods for solving IEP1, assuming the existence of a solution, and the modifications required for the multiple eigenvalue case. We shall not discuss conditions for the existence of a solution, which may be found in the literature for many special cases. Neither shall we discuss techniques for ensuring global convergence of the numerical methods, an important practical matter.

We should mention that although most inverse eigenvalue problems require iterative methods for their solution, there are some which can be solved by direct methods. One such case is the Jacobi matrix reconstruction problem; see [6] and [12].

Notation. Let $\{\lambda_i(c)\}_1^n$ be the eigenvalues of A(c) numbered in increasing order. Since A(c) is symmetric, it has an orthogonal set of eigenvectors $\{q_i(c)\}_1^n$ corresponding to $\{\lambda_i(c)\}_1^n$. Let $Q(c) = [q_1(c) \ldots q_n(c)]$ and let $\Lambda^* = \mathrm{Diag}(\lambda_i^*)$. Let c^* be a solution to Problem IEP1.

2. DISTINCT EIGENVALUES

In this section we assume that the given eigenvalues are distinct, i.e. $\lambda_1^* < \lambda_2^* < \ldots < \lambda_n^*$. Since c^* is a solution of IEP1 and the eigenvalues are continuous functions of c, it follows that $\{\lambda_i(c)\}_1^n$ are distinct in a neighborhood of c^*. They are therefore differentiable in a neighborhood of c^*, with J(c), the Jacobian of the function $\lambda(c)$, given by

$$J_{ik}(c) = \frac{\partial \lambda_i(c)}{\partial c_k} = q_i(c)^T A_k q_i(c) \tag{2.1}$$

(see [20] or [15]). We define Method I to be Newton's method applied to solve

$$f(c) = \lambda(c) - \lambda^* = 0 .$$

It consists of the iteration

$$J(c^\nu)(c^{\nu+1} - c^\nu) = \lambda^* - \lambda(c^\nu) .$$

Using the fact that A(c) is affine and that $(\lambda_i(c), q_i(c))$ is an eigenpair of A(c), this iteration may also be written as

$$q_i(c^\nu)^T A(c^{\nu+1}) q_i(c^\nu) = \lambda_i^* , \quad i = 1,\ldots,n .$$

Indeed, this last equation is the motivation behind the use of Method I by physicists in solving certain problems arising in spectroscopy (see [3]). This method has been discussed by many authors including Downing and Householder [7] and Kublanovskaja [16]. It is locally quadratically convergent assuming $J(c^*)$ is nonsingular.

Method I requires the computation of the eigenvectors $\{q_i(c^\nu)\}$ for every iterate c^ν. An alternative approach is to approximate the eigenvectors. We define Method II to consist of the iteration

$$(q_i^\nu)^T A(c^{\nu+1}) q_i^\nu = \lambda_i^* , \quad i = 1,\ldots,n \tag{2.2}$$

where each approximate eigenvector q_i^ν is obtained by taking one step of inverse iteration using the target eigenvalue λ_i^*, i.e.

$$\begin{aligned} (A(c^\nu) - \lambda_i^* I)\gamma_i &= q_i^{\nu-1} \\ q_i^\nu &= \gamma_i / \|\gamma_i\| . \end{aligned} \tag{2.3}$$

Note that (2.2) is simply a linear system with a coefficient matrix which approximates $J(c^\nu)$. Although only one step of inverse iteration is taken for each eigenvector, this is enough to make the overall iteration (2.2) quadratically convergent, assuming $J(c^*)$ is nonsingular. Method II is essentially due to Osborne [21] who used a different form of the iteration intended for nonlinear functions A(c).

Consider a third method motivated as follows. By solving IEP1 we mean finding c and Q such that

$$Q^T A(c) Q = \Lambda^* \tag{2.4}$$

Now suppose we have a pair of estimates c^ν, $Q^{(\nu)}$ and we wish to replace them by better estimates $c^{\nu+1}$, $Q^{(\nu+1)}$. Since $Q^{(\nu)}$, $Q^{(\nu+1)}$ are orthogonal matrices let us write

$$Q^{(\nu+1)} = Q^{(\nu)} e^{Y} \tag{2.5}$$

where Y is a skew-symmetric matrix, i.e. $Y = -Y^T$. Substituting $Q^{(\nu+1)}$ into (2.4) and expanding e^Y we find

$$(Q^{(\nu)})^T A(c) Q^{(\nu)} = \Lambda^* + Y\Lambda^* - \Lambda^* Y + O(\|Y\|^2) . \tag{2.6}$$

Define $c^{\nu+1}$ by neglecting second-order terms and equating diagonal elements; this gives precisely equation (2.2), where $Q^{(\nu)} = [q_1^\nu \ldots q_n^\nu]$. Now equate off-diagonal terms of (2.6), again neglecting second-order terms. This gives

$$(q_i^\nu)^T A(c^{\nu+1}) q_j^\nu = y_{ij}(\lambda_j^* - \lambda_i^*) , \quad 1 \le i < j \le n . \tag{2.7}$$

Since $\{\lambda_i^*\}_1^n$ are distinct, the skew-symmetric matrix Y is completely determined by (2.7). Forming e^Y would be very expensive, so we replace (2.5) by an approximation:

$$Q^{(\nu+1)} = Q^{(\nu)}(I + \tfrac{1}{2} Y)(I - \tfrac{1}{2} Y)^{-1} \tag{2.8}$$

This completely defines Method III, which, like Methods I and II, is locally quadratically convergent. Method III is apparently new.

We see that all three methods can be summarized as follows:

Methods I, II, III.

Choose c^0. Form $A(c^0)$ and compute its eigenvectors $Q^{(0)} = [q_1^0 \ldots q_n^0]$.

For $\nu = 0,1,2,\ldots$

Step 1. If $\|(Q^{(\nu)})^T A(c^\nu) Q^{(\nu)} - \Lambda^*\|_F$ is sufficiently small, stop.

Step 2. Form $J^{(\nu)}$ and b^ν where

$$J_{ik}^{(\nu)} = (q_i^\nu)^T A_k q_i^\nu ,$$

$$b_i^\nu = (q_i^\nu)^T A_0 q_i^\nu .$$

Obtain $c^{\nu+1}$ by solving

$$J^{(\nu)} c^{\nu+1} = \lambda^* - b^\nu.$$

Form $A(c^{\nu+1})$.

Step 3. (Definition of $Q^{(\nu+1)} = [q_1^{\nu+1} \ldots q_n^{\nu+1}]$).

Method I: Set $Q^{(\nu+1)} = Q(c^{\nu+1})$, the matrix of eigenvectors of $A(c^{\nu+1})$.

Method II: Obtain $Q^{(\nu+1)}$ from (2.3), replacing ν by $\nu+1$.

Method III. Obtain $Q^{(\nu+1)}$ from (2.7), (2.8).

We derive a fourth method using a different viewpoint. Consider the nonlinear system

$$g(c) = \begin{bmatrix} \det(A(c) - \lambda_1^* I) \\ \vdots \\ \det(A(c) - \lambda_n^* I) \end{bmatrix} = 0 .$$

We define Method IV to be Newton's method applied to solve this nonlinear system (see [1]). Using the fact that

$$g_i(c) = \prod_{k=1}^{n} (\lambda_k(c) - \lambda_i^*) ,$$

it is not difficult to show that Method IV consists of the same iteration as Method I, except that the matrix $J(c^\nu)$, defined by (2.1), must be premultiplied by $W(c^\nu)$, where

$$w_{ij}(c) = \frac{\lambda_i(c) - \lambda_i^*}{\lambda_j(c) - \lambda_i^*} .$$

Since the given eigenvalues $\{\lambda_i^*\}_i^n$ are distinct, $W(c) \to I$ as $c \to c^*$, and so asymptotically both methods coincide, both being quadratically convergent. Nonetheless, our numerical experience indicates that Method I almost always requires fewer iterations, and that Method IV suffers more often from ill-conditioning. It seems that Method I is always a better choice than Method IV for practical use.

The choice among Methods I, II, III is not so clear. All three are locally quadratically convergent assuming only that $J(c^*)$ is nonsingular. All three have very similar performance in practice, requiring almost exactly the same number of iterations to achieve a given accuracy in most of our tests. Method I has the theoretical disadvantage that Step 3 requires an iterative procedure to find the eigenvectors $\{q_i(c^{\nu+1})\}$. In practice, however, this task is considered to generally require only about $5n^3$ multiplication operations (see [11, p. 282]). A naive implementation of Step 3, Method II would require $O(n^4)$ operations. Instead, one should first perform a Householder tridiagonalization of $A(c^{\nu+1})$, requiring $\frac{2}{3}n^3$ multiplications (see [11] or [22]) and then solve n tridiagonal systems. Each tridiagonal system requires n^2 multiplications to obtain the transformed right-hand side, $O(n)$ operations to solve the system, and n^2 multiplications to transform the solution. The total number of multiplications required for Step 3 of Method II is therefore about $3n^3$.

Step 3 of Method III requires about $4n^3$ operations. Note also that if the $\{A_k\}$ are dense, n^4 operations are required by all the methods simply to form J in Step 2. However, in the case of the additive inverse problem (1.2), forming J requires only n^2 multiplications.

3. MULTIPLE EIGENVALUES

In this section we suppose that the set $\{\lambda_i^*\}_1^n$ includes multiple eigenvalues. For convenience we assume that there is only one multiple eigenvalue, with multiplicity t, so that

$$\lambda_1^* = \lambda_2^* = \dots = \lambda_t^* < \lambda_{t+1}^* < \dots < \lambda_n^* .$$

There is no difficulty in generalizing all our remarks to an arbitrary set of eigenvalues.

We first consider the behaviour of Method I. When all the eigenvalues $\{\lambda_i(c)\}$ are distinct, their first partial derivatives are given by J(c) (see (2.1)), and it is not difficult to show that their second partial derivatives are given by

$$\frac{\partial^2\lambda_i}{\partial c_k \partial c_j} = 2 \sum_{\substack{\ell=1 \\ \ell\neq i}}^{n} \frac{[q_\ell(c)^T A_k q_i(c)]\,[q_\ell(c)^T A_j q_i(c)]}{\lambda_i(c) - \lambda_\ell(c)} . \tag{3.1}$$

The quadratic convergence of Newton's method depends on the existence and Lipschitz continuity of the Jacobian in a neighbourhood of the solution, with the size of the region of attraction varying inversely in proportion to the size of the Lipschitz constant. Since the second partial derivatives (3.1) give a value for the Lipschitz constant for J(c), we see that as the separation of the eigenvalues decreases, the size of the region of attraction also decreases (the problem is more ill-conditioned). When the separation is zero, the eigenvalues $\{\lambda_i(c)\}$ are generally not differentiable at c^*, and the eigenvectors $\{q_i(c)\}$, which are not uniquely defined at c^*, generally cannot be defined as continuous functions of c at c^*. As long as the eigenvalues of $A(c^\nu)$ are distinct for all iterates c^ν, Method I remains well defined even when $\{\lambda_i^*\}$ contains multiple eigenvalues. However, the matrix of eigenvectors $Q(c^\nu)$, and consequently the Jacobian $J(c^\nu)$, generally will not converge as $c^\nu \to c^*$. Therefore one might expect that, at the very least, the quadratic convergence property of Method I will break down.

In fact, however, the convergence is generally quadratic, both in theory and in practice. Bohte (1967-68) mentions that in his numerical experiments, multiple eigenvalues caused no difficulties, but he gives no explanation for this. Although Method I has been used and discussed by many authors, its behaviour in the presence of multiple eigenvalues does not seem to have been addressed until recently, when Nocedal and Overton [18] gave a convergence analysis. This analysis is based on a classical result of Rellich [25], which states that the eigenvalues of a matrix which is an analytic function of a single variable can always be numbered so that they are each analytic functions of the variable. Similarly, the eigenvectors at points of multiple eigenvalues can always be defined so that they are continuous (indeed analytic) functions of the single variable. This result is applied to the eigenvalues of $A(c)$ considered along a single line passing through the solution c^* and the current iterate c^ν. By using the mean value theorem for one variable it follows that, locally, every Newton step produces a quadratic contraction in the error. The result is that, given a nonsingularity condition, the iterates c^ν converge quadratically although the sequence $\{J(c^\nu)\}$ does not converge. The situation is illustrated in Figure 1.

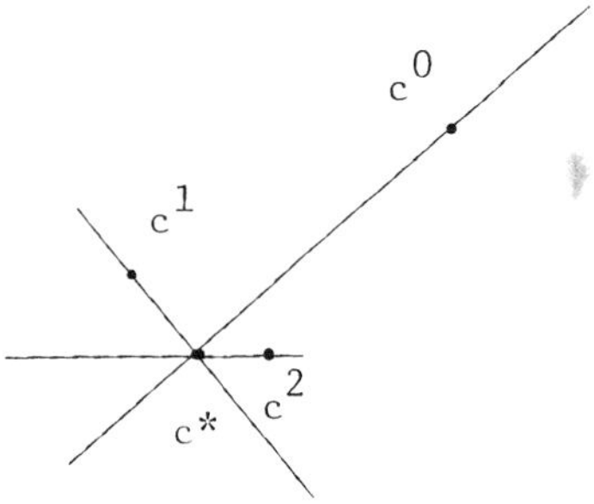

Figure 1. Convergence of Method I.

The eigenvalues are well-behaved differentiable functions along the lines shown in the illustration, all of which pass through c^*. It is on this fact that the analysis is based. By contrast, consider the eigenvalues along lines near c^* but not passing through this point. Generally they would be differentiable functions, but they would be very ill-conditioned. More specifically, $J(c)$ would vary extremely rapidly along these lines, since the eigenvalue separation is very small near c^*.

It is interesting to consider the nonsingularity assumption needed for the quadratic convergence proof. A sufficient condition is that $\{A(c^\nu)\}$ has distinct eigenvalues for all ν and that $\{\|J(c^\nu)^{-1}\|\}$ is bounded. Although this situation usually holds in practice, it would be more desirable to state the nonsingularity assumption in terms of a matrix evaluated at c^*. Since Q, the matrix of eigenvectors, is not uniquely defined at c^*, let us define

$$\Omega = \{Q: Q^TQ = I \text{ and } Q^TA(c^*)Q = \Lambda^*\}$$

and, for any $Q \in \Omega$, define $J^*(Q)$ by (2.1) using $Q = [q_1 \ldots q_n]$. We might hope that

$$\sup_{Q \in \Omega} \|J^*(Q)^{-1}\| < \infty .$$

However, it turns out that, in general, this supremum does not exist, and furthermore that, when $n \geq t(t-1)/2$, it is generally possible to find a point c which is near c^* and for which $J(c^*)$ is arbitrarily near singularity; see [9]. Nonetheless, it is interesting to note that even when Method I is started at such a point c^0, it usually has no difficulties with convergence. Typically the next iterate c^1 is such that $J(c^1)$ is not nearly singular. Although the error $\|c^1-c^*\|$ is not quadratic in $\|c^0 - c^*\|$, subsequent iterates converge quadratically as usual.

Before going on to Methods II and III, let us briefly consider Method IV in the multiple eigenvalue case. This analysis is trivial. The function g(c) is differentiable, but it has t identical components $g_1(c), \ldots, g_t(c)$, so the method must be reformulated. Since it seems an inferior method even in the distinct eigenvalue case, we do not consider it any further.

Let us consider Methods II and III. These have a similar behaviour to that of Method I. In particular, it is generally possible that $J^{(\nu)}$ could be arbitrarily near singularity for some c^ν near c^*, although again this is unlikely to happen. The implementation of Method II must be done with some care in order to obtain an independent set $\{q_i^\nu\}_1^t$ corresponding to the multiple eigenvalue. This may be done by explicitly orthogonalizing the vectors $\{q_i\}_1^t$ as they are obtained from the inverse iteration (see [23]). The behaviour of Method III is particularly interesting in the multiple eigenvalue case. The components of Y given by y_{ij}, $1 \leq i < j \leq t$, are not defined by (2.7) since the right-hand side is identically zero. We must therefore choose values for these y_{ij}. Because of the relation (2.5) (or its approximation (2.8)), it can be seen that y_{ij}, $1 \leq i < j \leq t$, correspond to the rotation of the current eigenvector iterates $[q_1^\nu \ldots q_t^\nu]$

within the corresponding invariant subspace to approximate the new eigenvector iterates $[q_1^{\nu+1} \ldots q_t^{\nu+1}]$. More specifically, suppose $t = 2$, and consider the plane P^ν which is spanned by q_1^ν, q_2^ν and the plane $P^{\nu+1}$ which is spanned by $q_1^{\nu+1}$, $q_2^{\nu+1}$. As $c^\nu \to c^*$, P^ν converges to P^*, the plane which is the invariant subspace corresponding to $\lambda_1^* = \lambda_2^*$. Of course, any orthogonal pair of eigenvectors in P^* is a pair of eigenvectors for $\lambda_1^* = \lambda_2^*$. Now y_{12} corresponds to the rotation of q_1^ν, q_2^ν within the plane P^ν to line these vectors up with $q_1^{\nu+1}$, $q_2^{\nu+1}$ respectively. A little thought makes it clear that the appropriate value which should be assigned to y_{12} is zero. A nonzero value amounts to unnecessarily causing the iterates (q_1^ν, q_2^ν) to spin around in the approximate invariant subspace as $\nu \to \infty$. Divergence of the eigenvector approximations $\{q_i^\nu\}$ is not a problem in itself; indeed, this exactly describes their behaviour in Methods I and II. However, Method III will not converge if y_{12} is set to too large a value, since $I + Y$ will not adequately approximate e^Y in (2.6). For a rigorous discussion, see [9]. There it is shown that when y_{ij} is set to zero, $1 \leq i < j \leq t$, not only does c^ν converge to c^* quadratically, as in Methods I and II, but also $Q^{(\nu)}$ converges quadratically to a limit, which is not the case for Methods I and II.

The conclusion drawn so far is that Methods I, II and III are generally locally quadratically convergent in the multiple eigenvalue case provided that a solution c^* exists, and provided that the $\{q_i^\nu\}_1^t$ are orthogonalized in Method II and the y_{ij}, $1 \leq i < j \leq t$, are set to zero in Method III. However, it is quite worthwhile to devote further thought to the behaviour of Method III. Consider again equation (2.7), with $1 \leq i < j \leq t$. Although we are setting $y_{ij} = 0$, we see that the equation

$$(q_i^\nu)^T A(c^{\nu+1}) q_j^\nu = 0 , \quad 1 \leq i < j \leq t \tag{3.2}$$

is not being imposed by the method. This set of $t(t-1)/2$ linear equations could be used as additional conditions which $c^{\nu+1}$ must satisfy. Since $c^{\nu+1}$, a vector of n parameters, is already constrained by the n conditions (2.2), we see that our problem is, in some sense, overdetermined. The same conclusion may be obtained more directly by considering equation (2.4). This appears to represent $n(n+1)/2$ equations with $n(n+1)/2$ degrees of freedom in c and Q, since an orthogonal matrix has $n(n-1)/2$ degrees of freedom. However, the $t(t-1)/2$ degrees of freedom in the choice of basis to represent the invariant subspace corresponding to the multiple eigenvalue are of no use in solving (2.4). Thus we see again that Problem IEP1 is essentially missing $t(t-1)/2$ degrees of freedom. Consequently, it is unreasonable to expect IEP1 to have a solution c^*. We must either remove some of the conditions demanded by IEP1 or introduce new parameters. For simplicity, we take the former course, and reformulate the problem as follows:

IEP2: Find $c \in \mathbb{R}^n$ such that $A(c)$, given by (1.1), has its first $n-s$ eigenvalues given by the prescribed values

$$\lambda_1^* = \lambda_2^* = \dots = \lambda_t^* < \lambda_{t+1}^* < \dots < \lambda_{n-s}^*$$

where $s = t(t-1)/2$.

The corresponding modification which must be made to Method III is now immediate. Since $s = t(t-1)/2$ of the eigenvalues are no longer known, the corresponding conditions are removed from (2.2) and replaced by (3.2), making n conditions in n unknowns. These n conditions may be rewritten as

$$(Q_1^{(\nu)})^T A(c^{\nu+1}) Q_1^{(\nu)} = \lambda_1^* I \qquad (3.3)$$

$$(q_i^\nu)^T A(c^{\nu+1}) q_i^\nu = \lambda_i^*, \quad i = t+1, \dots, n-s,$$

where $Q_1^{(\nu)} = [q_1^\nu, \dots, q_t^{(\nu)}]$. Provided the matrix of coefficients describing (3.3) is nonsingular in the limit, the modified method is quadratically convergent. Note that this nonsingularity condition is independent of the basis for $Q_1^{(\nu)}$, unlike before, where the singularity of $J^{(\nu)}$ depended on the choice of eigenvector basis. Thus the modified method has two advantages. The primary advantage is that the modified method is quadratically convergent with less data than the unmodified method. A secondary advantage is that for almost all problems, it is possible, though extremely unlikely, that the unmodified method could lose quadratic convergence if the trajectory of iterates $\{c^\nu\}$ is unfortunate. This is not the case for the modified method.

Methods I and II should also be modified in the same way, so that $c^{\nu+1}$ is defined by (3.3) instead of (2.2). They are quadratically convergent under the same nonsingularity condition just mentioned for Method III. The fact that $Q^{(\nu)}$ does not converge for Methods I and II is of little consequence.

With hindsight it can be seen that the modified version of Method I is essentially equivalent to Newton's method applied to solve:

$$f_1(c) = Q_1(c)^T A(c)\, Q_1(c) - \lambda_1^* I = 0$$

$$(f_2(c))_i = q_i(c)^T A(c)\, q_i(c) - \lambda_i^* = 0 , \quad i = t+1,\ldots,n-s$$

where $Q_1(c) = [q_1(c),\ldots,q_t(c)]$. See [9] for details.

To our knowledge, the modified methods are new. Indeed, we are not aware of any discussion of the correct general formulation of the inverse eigenvalue problem in the multiple eigenvalue case, choosing the dimension of the parameter space as in IEP2. However, the dimension argument is well known in other contexts, which we now briefly mention. First, a related concept is known in quantum mechanics as the "crossing rule" of von Neumann and Wigner; see [10]. Second, it is well known in the structural engineering literature that the dimension of the parameter space must be increased when multiple eigenvalues arise as active constraints in a minimization process; see [19]. Third, the dimension argument is well known in the context of the problem of communality (mentioned in Section 1) and the related minimization problem studied by Fletcher. Here there is one multiple eigenvalue equal to zero, and the dimension argument amounts to saying that

$$\lambda_1^* = \lambda_2^* = \ldots = \lambda_t^* = 0$$

is generally $t(t+1)/2$ independent conditions on c. For example, a specified nullity of $t = 3$ is generally enough to locally determine a solution c in a space of dimension $n = 6$. Harman [14, p. 69] obtains this formula by an argument involving rank and determinants, while Fletcher [8, p. 502] obtains it by setting a $t \times t$ block of a block Choleski factorization equal to zero.

We conclude this section by mentioning a few numerical methods which have been published to solve some of the applied problems discussed in Section 1. In the engineering literature there are various methods for solving optimization problems with eigenvalue constraints. Since the constraints are nondifferentiable in the multiple eigenvalue case, some of these methods use subgradients. See particularly Polak and Wardi [24]; see also Cullum, Donath and Wolfe [5], who solve a related problem. These methods are first-order. Choi and Haug [4] give an algorithm to solve an optimal design problem with a constraint involving one double eigenvalue; this method is apparently quadratically convergent. We have in mind that, in order to obtain quadratic convergence, a general minimization method should be based on the modified methods for IEP2; in particular, it should incorporate condition (3.2). Of course, one must also introduce Lagrange multipliers and the appropriate second derivatives. At present we have not done this, confining our attention to the model problem IEP2.

Harman [14] gives several methods to solve the problem of communality, but apparently none of these is second-order. The algorithm of Fletcher [8] is quadratically convergent and is in some ways quite closely related to our modified methods. However, its application is restricted to minimization problems with positive semi-definite constraints, i.e. nonnegativity constraints on the eigenvalues of A(c).

4. NUMERICAL RESULTS

We have tested Methods I, II, III and IV on many different test problems of various origins. We give here some detailed output of the modified versions of Methods I, II, III for one simple example of small dimension. Results for some larger, more difficult, problems are reported in [9]. The results were obtained using a Vax 11/780 at the Courant Mathematics and Computing Laboratory, using double precision arithmetic (approximately 16 decimal digits of accuracy). In the following we write vectors as row-vectors for convenience.

The test example is an additive inverse eigenvalue problem, defined by (1.2) and

$$A_0 = VV^T$$

where V is the 4×2 matrix

$$\begin{bmatrix} 1.4 & 1.9 \\ 0.3 & 1.2 \\ 1.7 & 9.1 \\ 2.4 & 0.1 \end{bmatrix}$$

Now let $c^0 = (1,1,1,1)$. Then

$$A(c^0) = VV^T + I ,$$

so that by construction $A(c^0)$ has one double eigenvalue with value one. In fact the eigenvalues of $A(c^0)$ are, to six figures,

$$\lambda(c^0) = (1,\ 1,\ 7.39427,\ 93.1757).$$

Since the dimension n is equal to 4, a suitable configuration of target eigenvalues is as follows: specify one double eigenvalue (three conditions) and one single eigenvalue (one condition), i.e. $t = 2$, $s = 1$, $n-s = 3$ in IEP2.

If we were to specify the target eigenvalues $\lambda^* = (1,1,7.39427)$ there would be one locally unique solution $c^* = c^0$. Instead we choose

$$\lambda^* = (1.3,\ 1.3,\ 7.0)\ .$$

The output of modified Methods I, II and III, using c^0 as a starting point, is shown in Table I. In all cases the iterates c^ν converge quadratically to the same locally unique solution c^*. Only 3 iterations are required for each method to achieve full accuracy since c^0 is so near c^*. The first row of the matrix $Q^{(\nu)}$ is shown at each iteration; this gives the first component of the eigenvectors of $A(c^\nu)$ in the case of Method I and the first component of their approximations in the case of Methods II and III. Note that the components corresponding to the double eigenvalue do not converge for Methods I and II but that they do converge quadratically for Method III. The Rayleigh quotients printed are the quantities

$$(q_i^\nu)^T A(c^\nu) q_i^\nu\ ;$$

these are the eigenvalues of $A(c^\nu)$ in the case of Method I and approximations to these in the case of Methods II and III. Note that the fourth Rayleigh quotient converges to the eigenvalue 94.027 although this value is not specified by λ^*. The residual norm shown is

$$\| (\bar{Q}^{(\nu)})^T A(c^\nu) \bar{Q}^{(\nu)} - \Lambda^* \|_F$$

where $\bar{Q}^{(\nu)}$ consists of the first three columns of $Q^{(\nu)}$ and $\Lambda^* = \mathrm{Diag}(1.3,\ 1.3,\ 7.0)$.

We used the NAG library to compute the eigenvectors required at each step of Methods II and III. (We could equally well have used EISPACK.) The initial vectors q_1^0, q_2^0 are chosen arbitarily by the subroutine, since $\lambda_1(c^0) = \lambda_2(c^0)$. The LINPACK library was used to solve the various linear systems required.

```
OUTPUT OF MODIFIED METHOD I
   First row of Q is    -0.8978591    0.0718796   -0.3712337   -0.2255389
   Rayleigh quotients are    1.0000000    1.0000000    7.3942653   93.1757347
       Residual norm =           5.79E-01
  Iteration 1. c =    1.4476239    1.2930937    1.7499035    0.4363120
   First row of Q is    -0.8709080    0.0814436   -0.4293405   -0.2248397
   Rayleigh quotients are    1.2754403    1.2999696    7.0244217   93.8971015
       Residual norm =           3.46E-02
  Iteration 2. c =    1.4863409    1.2910892    1.8860821    0.3918019
   First row of Q is    -0.8646268    0.1131475   -0.4349317   -0.2246165
   Rayleigh quotients are    1.2997649    1.2999975    7.0002279   94.0253237
       Residual norm =           3.27E-04
  Iteration 3. c =    1.4867027    1.2910674    1.8877904    0.3913661
   First row of Q is    -0.8611041    0.1372387   -0.4349875   -0.2246134
   Rayleigh quotients are    1.3000000    1.3000000    7.0000000   94.0269267
       Residual norm =           3.27E-08
OUTPUT OF MODIFIED METHOD II
   First row of Q is    -0.8978591    0.0718796   -0.3712337   -0.2255389
   Rayleigh quotients are    1.0000000    1.0000000    7.3942653   93.1757347
       Residual norm =           5.79E-01
  Iteration 1. c =    1.4476239    1.2930937    1.7499035    0.4363120
   First row of Q is    -0.1797791   -0.8559108   -0.4295837   -0.2255389
   Rayleigh quotients are    1.2996534    1.2757569    7.0244213   93.8969330
       Residual norm =           3.47E-02
  Iteration 2. c =    1.4863720    1.2910874    1.8862052    0.3917653
   First row of Q is    -0.1158115   -0.8642716   -0.4349365   -0.2255389
   Rayleigh quotients are    1.2999975    1.2997846    7.0002084   94.0251860
       Residual norm =           3.00E-04
  Iteration 3. c =    1.4867027    1.2910674    1.8877905    0.3913661
   First row of Q is    -0.1384818   -0.8609050   -0.4349875   -0.2255389
   Rayleigh quotients are    1.3000000    1.3000000    7.0000000   94.0266721
       Residual norm =           2.73E-08
OUTPUT OF MODIFIED METHOD III
   First row of Q is    -0.8978591    0.0718796   -0.3712337   -0.2255389
   Rayleigh quotients are    1.0000000    1.0000000    7.3942653   93.1757347
       Residual norm =           5.79E-01
  Iteration 1. c =    1.4476239    1.2930937    1.7499035    0.4363120
   First row of Q is    -0.8715932    0.0720206   -0.4296476   -0.2248136
   Rayleigh quotients are    1.2754440    1.2999667    7.0244210   93.8971013
       Residual norm =           3.48E-02
  Iteration 2. c =    1.4863800    1.2910863    1.8862843    0.3917534
   First row of Q is    -0.8690117    0.0720770   -0.4349384   -0.2246157
   Rayleigh quotients are    1.2997911    1.2999970    7.0002027   94.0255133
       Residual norm =           2.92E-04
  Iteration 3. c =    1.4867027    1.2910674    1.8877905    0.3913661
   First row of Q is    -0.8689876    0.0720777   -0.4349875   -0.2246134
   Rayleigh quotients are    1.3000000    1.3000000    7.0000000   94.0269267
       Residual norm =           2.49E-08
```

TABLE I. Numerical Results for Modified Methods I, II, III

As a second experiment we tried splitting the specified double eigenvalue into two different target eigenvalues. This requires also specifying the fourth eigenvalue since otherwise there are not enough conditions to uniquely define a solution. We tried

$$\lambda^* = (0.99,\ 1.01,\ 7.3,\ 93.0)\ .$$

Methods I, II and III all converged to a solution

$$c^* = (1.019382,\ 1.011377,\ 0.809888,\ 0.889354)$$

without any difficulty. However, when the first two target eigenvalues were set closer together, as in

$$\lambda^* = (0.9999,\ 1.0001,\ 7.3,\ 93.0)\ ,$$

no convergence was obtained. There is no easy way of telling whether this problem has a solution; even if it does, the problem is very ill-conditioned so one cannot expect to find a solution easily. Certainly one cannot expect a solution to exist when

$$\lambda^* = (1,\ 1,\ 7.3,\ 93.0)\ ,$$

since then too many conditions are being imposed.

We have constructed some examples where almost multiple target eigenvalues are specified and where convergence does not take place except when the distance $\|c^0 - c^*\|$ is substantially smaller than the eigenvalue separation size. In these examples we know the solution c^* by construction of the problem. Such a case may be obtained by modifying Example 2 of [18]. In that example, a problem is defined with $n = 6$ and with a specified multiple eigenvalue of multiplicity 3. The solution is $c^* = (1,1,1,1,1,1)$. Now change c^* to (1.000001, 1.000001, 1.000001, 0.999999, 0.999999, 0.999999), and change λ^* accordingly. The target eigenvalues are then distinct with separation about 10^{-6}. When c^0 is set to (1,1,1,1,1,1), Method I fails to converge. However, if c^0 is set to within 10^{-9} of c^*, quadratic convergence takes place. By contrast, when modified Method I is applied to the original problem, specifying only the multiple eigenvalue

$$\lambda^* = (1,1,1)$$

and not the other eigenvalues, convergence rapidly takes place from the starting point $c^0 = (.9,\ .9,\ .9,\ 1.1,\ 1.1,\ 1.1)$.

In summary we have seen that inverse eigenvalue problems with multiple eigenvalues can be solved very efficiently when formulated properly as in IEP2 and when the methods are modified accordingly. This seems to be very important since multiple eigenvalues arise naturally in the constrained optimization context mentioned earlier. Problems where almost multiple eigenvalues are specified can be much more difficult to solve than those with multiple eigenvalues. However, we know of no practical applications where such ill-posed problems arise.

Acknowledgements. We would like to thank Gene Golub for making many helpful comments during the course of this work and bringing several references to our attention. We would also like to thank Olof Widlund for many useful discussions and Robert Kohn for introducing us to the relevant structural engineering literature. The work of the first author was supported in part by National Science Foundation Grant No. MCS 83-00842. The work of the second author was supported in part by National Science Foundation Grant No. DCR-8401903. The work of the third author was supported in part by National Science Foundation Grant Nos. DCR-8302021 and DCR-8502014.

REFERENCES

1. BIEGLER-KONIG, F. W. A Newton iteration process for inverse eigenvalue problems, Numer. Math. 37 (1981) 349-354.
2. BOHTE, Z. Numerical solution of the inverse algebraic eigenvalue problem, Comp. J. 10 (1967-68) 385-388.
3. BRUSSARD, P. J. and GLAUDEMANS, P. W. Shell Model Applications in Nuclear Spectroscopy, Elsevier, New York, 1977.
4. CHOI, K. K. and HAUG, E. J. A numerical method for optimization of structures with repeated eigenvalues, in: Optimization of Distributed Parameter Structures, Vol. I (eds. E. J. Haug and J. Cea), Sijthoff and Noordhoff, Alphen aan den Rijn, Netherlands, 1981, 534-551.
5. CULLUM, J., DONATH, W. E. and WOLFE, P. The minimization of certain nondifferentiable sums of eigenvalues of symmetric matrices, Math. Programming Study 3 (1975) 35-55.
6. de BOOR, C. and GOLUB, G. H. The numerically stable reconstruction of a Jacobi matrix from spectral data, Linear Algebra and its Appl. 21 (1978) 245-260.
7. DOWNING, A. C. and HOUSEHOLDER, A. S. Some inverse characteristic value problems, J. Assoc. Comput. Mach. 3 (1956) 203-207.
8. FLETCHER, R. Semi-definite matrix constraints in optimization, SIAM J. Control Optim. 23 (1985) 493-513.
9. FRIEDLAND, S., NOCEDAL, J. and OVERTON, M. L. The formulation and analysis of numerical methods for inverse eigenvalue problems,

Report 8503-NAM-03, Dept. Comp. Sci. & Elec. Engr., Northwestern University, Evanston, Illinois, 1985.

10. FRIEDLAND, S., ROBBIN, J. W. and SYLVESTER, J. H. On the crossing rule, Comm. Pure Appl. Math. 37 (1984) 19-37.

11. GOLUB, G. H. and van Loan, C. Matrix Computations, Johns Hopkins University Press, Baltimore, 1983.

12. GRAGG, W. and HARROD, W. J. The numerically stable reconstruction of Jacobi matrices from spectral data, Report, University of Kentucky, 1984.

13. HALD, O. On discrete and numerical Sturm-Liouville problems, Ph.D. dissertation, Dept. of Math., New York Univ., 1972.

14. HARMAN, H. Modern Factor Analysis, University of Chicago Press, Chicago, 1967.

15. KATO, T. Perturbation Theory for Linear Operators, Springer-Verlag, New York, 1966.

16. KUBLANOVSKAJA, W. N. On an approach to the solution of the inverse eigenvalue problem, Zapiski Nauk. Sem. Lenin. Otdel. Math. Inst. in V. A. Steklova Akad. Nauk SSSR, 1970 (in Russian) 138-149.

17. MAYNE, D. Q. and POLAK, E. Algorithms for the design of control systems subject to singular value inequalities, Math. Programming Study 18 (1982) 112-134.

18. NOCEDAL, J. and OVERTON, M. L. Numerical methods for solving inverse eigenvalue problems, in: Numerical Methods (eds. V. Pereyra and A. Reinoza), Lecture Notes in Mathematics 1005, Springer Verlag, New York, Heidelberg and Berlin, 1983, 212-226.

19. OLHOFF, N. and TAYLOR, J. E. On structural optimization, J. Appl. Mech. 50 (1983) 1138-1151.

20. ORTEGA, J. M. Numerical Analysis, Academic Press, New York and London, 1972.

21. OSBORNE, M. R. On the inverse eigenvalue problem for matrices and related problems for difference and differential equations, in: Conference on Applications of Numerical Analysis (ed. J. Ll. Morris), Lecture Notes in Mathematics 228, Springer Verlag, New York, Heidelberg, and Berlin, 1971, 155-168.

22. PARLETT, B. N. The Symmetric Eigenvalue Problem, Prentice-Hall, Englewood Cliffs, New Jersey, 1981.

23. PETERS, G. and WILKINSON, J. H. The calculation of specified eigenvectors by inverse iteration, in: Handbook for Automatic Computation, Vol. II: Linear Algebra (eds. J. H. Wilkinson and C. Reinsch), Springer Verlag, New York, Heidelberg and Berlin, 1971, 418-439.

24. POLAK, E. and WARDI, Y. Nondifferentiable optimization algorithm for designing control systems having singular value inequalities, Automatica 18 (1982) 267-283.

25. RELLICH, F. Perturbation Theory of Eigenvalue Problems, Gordon and Breach, New York, 1969.

S. Friedland
Department of Mathematics, Statistics
and Computer Science
University of Illinois at Chicago
Chicago, Illinois 60680, USA

J. Nocedal
Department of Electrical Engineering
and Computer Science
Northwestern University
Evanston, Illinois 60201, USA

M. L. Overton
Courant Institute of Mathematical Sciences
New York University
251 Mercer Street
New York, New York 10012, USA

D GOLDFARB

Strategies for constraint deletion in active set algorithms

1. INTRODUCTION

Active-set feasible point methods are widely used for solving the linearly constrained optimization problem:

$$\begin{aligned} &\underset{x \in R^n}{\text{minimize}} \quad f(x) \\ &\text{subject to} \quad C^T x \geq b \end{aligned} \qquad (1)$$

where f(x) is a sufficiently smooth and, in general, nonlinear function of the n-vector x, C^T is an m x n matrix, b is an m-vector, and superscript T denotes transposition.

In this paper we give strategies for simultaneously dropping more than one constraint from the active set A which ensure that the step taken after the constraints are dropped does not violate any of the constraints in A. We also consider the criterion for choosing a particular constraint to drop when there are several possibilities from which to choose.

The outline of the rest of this paper is as follows. In the next section we discuss the basis for choosing descent directions in feasible point active set methods and describe two extreme strategies for deleting constraints from the active set. Our main results are presented in section 3. There we derive recurrence formulas which allow us to efficiently check whether several constraints can be simultaneously deleted from the active set. In section 4 we give two dropping strategies based upon these results and discuss their implementation. In section 5 we illustrate the use of one of these strategies in a primal active-set quadratic programming algorithm. In section 6 we present a "steepest-face" constraint deletion rule for choosing amongst several candidates and show how to implement it in the algorithms presented in section 5. Finally, we give an example in an appendix which illustrates certain aspects of our strategies.

* This research was supported in part by the National Science Foundation under Grants DCR-83-41408 and CDR-84-21402 and in part by the Army Research Office under Contract No. DG29-83-0106.

2. BASIC IDEA AND PRELIMINARIES

In active-set feasible-point methods all iterates are feasible; i.e., they satisfy all of the linear inequality constraints in (1). We refer to those constraints in (1) that are satisfied as equalities at the current iterate x as "tight" constraints and we define the active set to be some linearly independent subset of the tight constraints. (Some authors use "active" to mean "tight" and use "working set" to indicate what we refer to as the "active set.") We shall use K to denote the set $\{1,2,\ldots,m\}$ of indices of the constraints in (1) and A and E ($A \subseteq E \subseteq K$) to denote, the sets of indices of the active and tight constraints, respectively.

In active-set feasible-point methods the direction of search at each iteration is chosen to be a feasible descent direction p for the problem

$$\begin{array}{ll} \underset{p \in R^n}{\text{minimize}} & f(x + p) \\ \text{subject to} & n_i^T p \geq 0, \quad i \in A \end{array}$$

where n_i is the i-th column of C. In most algorithms of this type the direction p is obtained by replacing f(x+p) by a local quadratic approximation q(p) at the point x and solving the equality constrained quadratic program

$$\begin{array}{ll} \underset{p \in R^n}{\text{minimize}} & q(p) \\ \text{subject to} & n_i^T p = 0, \quad i \in \bar{A} \subseteq A \end{array}$$

obtained by requiring a subset $\bar{A}$ of the active constraints to be satisfied as equalities, and ignoring the rest. The quadratic approximation q(p) usually has the form $q(p) = 1/2 p^T G p + g^T p + q_0$, where $q_0 = f(x)$, $g = \nabla f(x)$ and $G = \nabla^2 f(x)$ or some approximation to the Hessian of f(x) as in quasi-Newton methods. Some methods like the gradient projection method [13], use a linear approximation to f(x+p) rather than a quadratic. If problem (1) is a quadratic program, then, of course, q(p) = f(x+p).

Our concern in this paper is how to choose the reduced active set $\bar{A}$, and hence, search direction p. First let us describe two extreme strategies: one which chooses the best feasible descent direction and the other which chooses a feasible descent direction with a minimal amount of work.

Let the active set at the current iterate x consist of the first k constraints of K and denote it by A_k. If A_k=E, then, with respect to the quadratic model, the best feasible search direction, p, at x, is given by the solution of the quadratic program

$$\underset{p\in R^n}{\text{minimize}} \quad Q(p) = q(x+p) - q(x) = 1/2p^TGp + g^Tp \tag{2a}$$

$$\text{subject to} \qquad N_k^Tp \geq 0. \tag{2b}$$

where $N_k = [n_1,\ldots,n_n]$.

If G is a positive definite symmetric matrix then p is the optimal solution to the quadratic program (2) if and only if the Kuhn-Tucker conditions, (2b),

$$\begin{bmatrix} G & -N_k \\ N_k^T & 0 \end{bmatrix} \begin{pmatrix} p \\ \alpha \end{pmatrix} = \begin{pmatrix} -g \\ 0 \end{pmatrix} , \tag{3a}$$

and

$$\alpha \geq 0 \tag{3b}$$

are satisfied. Since the normals to the constraints in the active set are assumed to be linearly independent, the operators

$$N_k^* = (N_k^TG^{-1}N_k)^{-1} \; N_k^TG^{-1} \tag{4}$$

and

$$H_k = G^{-1}(I-N_kN_k^*) \tag{5}$$

can be defined, and it is easy to verify that

$$\begin{bmatrix} G & -N_k \\ N_k^T & 0 \end{bmatrix}^{-1} = \begin{bmatrix} H_k & N_k^{*T} \\ -N_k^* & (N_k^TG^{-1}N_k)^{-1} \end{bmatrix} . \tag{6}$$

Consequently, if the vector of Lagrange multipliers is nonnegative, i.e.,

$$\alpha^k = N_k^*g \geq 0 , \tag{7}$$

then

$$p = -H_kg \tag{8}$$

is the optimal solution to the search direction problem (2). If $\alpha^k \not\geq 0$, then it is well-known that there is some subset A_q of the active set A_k containing $q < k$ constraints such that

$$p = -H_qg$$

is the optimal solution to (2), where H_q is defined by (4) and (5), and N_q is obtained from N_k by deleting all but the appropriate q columns of N_k. For this "reduced" active set A_q, the corresponding vector of Lagrange multipliers, $\alpha^q = N_q^*g$ is nonnegative.

Finding this optimal active set is, of course, equivalent to solving the quadratic program (2). If G is a positive definite symmetric matrix, which we shall henceforth assume, then we note that this problem is a "least-distance programming" problem. To see this we write Q(p) in (2) as

$$Q(p) = 1/2(p+G^{-1}g)^T G(p+G^{-1}g) - 1/2g^T G^{-1}g.$$

Thus, problem (2) can be stated as: find the vector p in the polyhedral cone $\mathcal{C} = \{p \in R^n | N^T p \geq 0\}$ that is "closest" to the Newton direction $-G^{-1}g$, where the distance between two vectors x and y is given by the G-norm

$$||x-y||_G = (x-y)^T G(x-y).$$

By applying the transformation

$$y = L^T(p+G^{-1}g)$$

where $G = LL^T$ this problem becomes one of finding in the space of the y variables the closest point in the transformed cone $\mathcal{C}$ to the origin in the ordinary Euclidean sense.

Although the above least distance programming problem is a special type of positive definite quadratic program, its solution can still require a substantial amount of computational effort.

If on the other hand, all that is required is a feasible descent direction, the computational task is much simpler. If $g \neq N_k u$, then $p = -H_k g$ gives such a direction. If $g = N_k u$ but $\alpha^k \equiv u \not\geq 0$, then

$$p = -H_{k-1}g \tag{9}$$

is a feasible descent direction, assuming that $\alpha_k^k < 0$, where α_i^k is the i-th component of α^k.

In between these to two extreme strategies -- i.e., dropping as many constraints as possible by solving a least distance programming problem versus dropping as few constraints as necessary and moving in the direction (8) or (9) -- there is a whole range of strategies that can be followed. We shall explore several of these with the goal of obtaining a direction that is close to the best search direction at a moderate computational cost.

3. MAIN RESULTS

We shall now derive conditions which ensure that when we drop several constraints from the active set A_k the direction of search

$$p = -H_q g$$

based upon the smaller active set A_q is feasible. Because H_q is positive semidefinite and

$$H_q w = 0 \quad \text{if, and only if,} \quad w = N_q v, \tag{10}$$

p will be a direction of descent as long as $g \neq N_q v$. In what follows we shall assume that we are at a point where the "starting" active set A_k includes all tight constraints; i.e., $A_k = E$. Further, we shall assume for notational convenience that whenever a constraint is dropped from the active set it is always the one with the highest index so that subsequent active sets A_q, $q < k$, consist of the first q constraints (i.e., $A_q = \{1,2,\ldots,q\}$).

Our first result is a set of recurrence formulas for the quantities

$$b_j^q = n_j^T H_q g, \quad q < j \leq k \tag{11}$$

which enables us to determine whether dropping the q-th constraint from A_q will cause the new step direction, $p = -H_{q-1} g$ to violate any previously dropped constraints.

Recurrence Relations for b_j^q:

$$b_q^{q-1} = \alpha_q^q \, n_q^T \, H_{q-1} \, n_q \tag{12a}$$

$$b_j^{q-1} = b_j^q + c_j^q \, b_q^{q-1} \quad , \quad q < j \leq k, \tag{12b}$$

where

$$c_j^q = n_j^T \, n_q^* , \tag{13}$$

$n_q^* = N_q^{*T} e_q$ and e_q is the q-th column of the identity matrix.

To derive these formulas we make use of the well-known recurrences for the operator H_q (e.g., see [1] or [4]),

$$H_{q-1} = H_q + \frac{H_{q-1} \, n_q \, n_q^T \, H_{q-1}}{n_q^T \, H_{q-1} \, n_q} = H_q + \frac{n_q^* \, n_q^{*T}}{n_q^{*T} \, G \, n_q^*} . \tag{14}$$

It follows from these recurrences that

$$H_{q-1} \, n_q = n_q^* / (n_q^{*T} \, G \, n_q^*) \tag{15}$$

and

$$n_q^T \, H_{q-1} \, n_q = 1/(n_q^{*T} \, G \, n_q^*), \tag{16}$$

and hence that

$$c_j^q = n_j^T \, H_{q-1} \, n_q / (n_q^T \, H_{q-1} \, n_q), \quad q < j \leq k \tag{17}$$

Multiplying (14) on the right by g, using (16), and the definitions of α^q and n_q^* yields

$$H_{q-1}g = H_q g + \alpha_q^q (n_q^T H_{q-1} n_q) n_q^* . \tag{18}$$

If this is then multiplied on the left by n_j, $q \le j \le k$, one obtains the recurrences (12) after noting that $H_q^T n_q = 0$.

If these recurrences are used, the work needed to compute b_j^q, $q < j \le k$, primarily results from the computation of c_j^q, $q < j \le k$. We now show how these quantities can be computed efficiently.

<u>Lemma 1</u>: $\sum_{i=q}^{j} \alpha_i^j c_i^q = \alpha_q^q$, $q \le j \le k$

<u>Proof</u>: From the definitions of H_j and α^j (see (5) and (7))

$$G^{-1}g = (H_j + G^{-1}N_j N_j^*)g = H_j g + G^{-1} N_j \alpha^j .$$

Premultiplying the above by $n_q^T P_{q-1}$, where

$$P_{q-1} = (I - N_{q-1} N_{q-1}^*)^T$$

and $q \le j \le k$, we obtain

$$b_q^{q-1} = n_q^T H_{q-1} g = \sum_{i=q}^{j} \alpha_i^j n_q^T H_{q-1} n_j \tag{19}$$

after using the fact that $P_{q-1} H_j g = H_j g$ and (10). It then follows from (12a) and (17) that (19) is equivalent to

$$\alpha_q^q n_q^T H_{q-1} n_q = \sum_{i=q}^{j} \alpha_i^j c_i^q n_q^T H_{q-1} n_q .$$

Lemma 1 then follows upon cancellation of $n_q^T H_{q-1} n_q$ from both sides of the last expression.

If t constraints have been dropped without taking a step where t=k-q, we see that computing $c_{q+1}^q, c_{q+2}^q, \ldots, c_k^q$, by solving the triangular system of Lemma 1 requires approximately $t^2/2$ operations compared with nt operations required by (13). An operation is either an addition or a subtraction and either a multiplication or a division. This assumes that the Lagrange multipliers α_i^j, $q \le i \le j \le k$ and the operators H_q and N_q^* have been stored. We note that in most implementations these operators are not explicitly available and computing n_q^* for use in (13) can require as much as an additional $O(n^2)$ operations.

No matter how the c_j^q are computed, we need the quantity $n_q^T H_{q-1} n_q$ to compute the b_j^q by the above recurrence relations. It costs as much as $O(n^2)$ operations to compute the reciprocal quantity $n_q^{*T} G n_q^*$. Alternately, as we shall show in section 6, the diagonal elements of the matrix $N_q^* G N_q^{*T} = (N_q^T G^{-1} N_q)^{-1}$ can be recurred as the active set changes.

Computing b_j^{q-1}, $j=q+1,\ldots,k$ from (12b) requires only an additional t operations.

The quantities b_j^{q-1}, $j=q+1,\ldots,k$ can also be computed directly from definition (11). In addition to requiring nt operations to perform the required t inner products, the tentative new search direction vector $H_{q-1}g$ is needed. This can be computed without first updating the operator, H_q, or its representation by using (18).

The conditions $b_j^{q-1} \leq 0$, $j=q,\ldots,k$ are clearly necessary and sufficient for the direction $-H_{q-1}g$ to be feasible. We now derive sufficient conditions which are much simpler (i.e., less expensive) to check. They are based upon the following recurrence formula for c_j^q which is valid as long as $\alpha_j^j \neq 0$ for $j=q,q+1,\ldots,k$.

Recurrence Relations for c_j^q:

$$c_j^q = \begin{cases} (1/\alpha_j^j) \sum_{i=q}^{j-1} (\alpha_i^{j-1} - \alpha_i^j) c_i^q , & q < j \leq k \\ 1 , & q = j \leq k \end{cases}$$

This recurrence relation can be easily derived by subtracting the relation in Lemma 1 from the same relation with j replaced by j-1, i.e.,

$$0 = \sum_{i=q}^{j-1} \alpha_i^{j-1} c_i^q - \sum_{i=q}^{j} \alpha_i^j c_i^q = -\alpha_j^j c_j^q + \sum_{i=q}^{j-1} (\alpha_i^{j-1} - \alpha_i^j) c_i^q.$$

It now follows from the recurrence relations for b_j^q and c_j^q that

Theorem 1:

If $A_k = E$,

$\alpha_i^{j-1} \leq \alpha_i^j$, $\quad q < j \leq k$,

$\quad q \leq i < j$

and

$$\alpha_j^j < 0, \qquad q \leq j \leq k,$$

then the direction $-H_{q-1}\, g$ is feasible.

The sufficient conditions of this theorem were first given in [4] for use in a quadratic programming algorithm. The recurrence relations for c_j^q and the triangular system of Lemma (1) that can be used to solve for these quantities was also given in [4]; however the proofs provided there are more complicated than those presented here. The recurrence relations for b_j^q have not been published before, although they were implemented, as suggested by the author, by Idnani in his thesis [11].

4. TWO CONSTRAINT DELETION STRATEGIES

The sufficient conditions of Theorem 1 allow one to very easily check if dropping a constraint whose current Lagrange multiplier, α_j^q, is negative will give a feasible direction $p = -H_{q-1}\, g$. This suggests the following strategies for constraint deletion. Let

$$J_1^q = \{j \mid \alpha_j^q < 0,\ j=1,\dots,q\},\quad J_2^q = \{j \mid \alpha_j^i \leq \alpha_j^{i+1},\ {i=q,\dots,k-1 \atop j=1,\dots,q}\},\quad \text{if } q < k,$$

and

$$J^q = \begin{cases} J_1^q, & \text{if } q = k \\ J_1^q \cap J_2^q, & \text{if } q < k \end{cases}$$

Note that $J_2^{q-1} \subseteq J_2^q$ whereas the sets J_1^{q-1} and J_1^q may not be nested. Recall that we are assuming that the constraint dropped from A_q is q.

Strategy 1:

(i) If $J^q \neq \emptyset$, choose a constraint $j \in J^q$ to drop.

(ii) Otherwise, end deletions.

Strategy 2:

(i) same as Strategy 1.

(ii) Otherwise (i.e., $J^q = \emptyset$), if $J_1^q \neq \emptyset$ choose a constraint $j \in J_1^q$ to drop if doing so results in all $b_j^{q-1} \leq 0$; otherwise end deletions.

To implement strategy 2 one would not recur the quantities b_j^q, $q < j \leq k$, until $J^q = \emptyset$. At that point one would compute them from their definition (11) in k-q inner products, using recurrence relations (12) thereafter. As indicated previously, if the set J_1^q has many elements then checking the b_j^{q-1} for all possible candidates can get very expensive.

It is important to stress several points about the above strategies. First, the order in which constraints are dropped determines the total number that can be dropped. The strategies discussed here are local and do not try to find the optimal order. To do so would require an amount of computation which increases exponentially with the total number of dropped constraints.

Second, our multiple dropping strategies do not result in any computational savings in updating the operators H_q and N_q^*, or their factorized representations, over performing the same active set changes over several steps. Savings, however, do usually occur in that these strategies result in fewer active set changes and fewer steps.

Third, our strategies allow us to check the feasibility of changing the active set without actually making that change (i.e., updating any operators). Consequently, that change does not have to be undone if the proposed search direction is infeasible. This can save a fair amount of wasted computation that could occur if a naive approach were followed.

Fourth, the relative efficiency of our strategies depends upon the particular algorithm and implementation in which it is imbedded. For example, strategy 1 can be implemented very efficiently within a primal quadratic programming algorithm as will be shown in the next section. There we use the Cholesky factorization

$$G = LL^T$$

and the QR factorization

$$L^{-1}N_q = Q\begin{bmatrix} R \\ 0 \end{bmatrix}$$

where L is (n x n) lower triangular, R is (q x q) upper triangular, and Q is an orthogonal matrix.

Instead of computing N_q^* and H_q explicitly, we store the matrices R and $J = L^{-T}Q$ and update them after each change in the active set. It is easily verified that

$$N_q^* = R^{-1}J_1^T$$

and

$$H_q = J_2J_2^T$$

where J_1 and J_2 consist of the first q and last n-q columns, respectively, of J.

The use of the above factorizations in quadratic programming was first suggested by Golub and Saunders [9] and, subsequently, employed in specific algorithms given in [6] and [7]. Their use in variable metric (quasi-Newton) active set algorithms for linearly constrained nonlinear programming is described in [5], where methods for updating these factorizations when the variable metric matrix G is changed, as well as when the active set is changed, are given.

Another commonly used implementation employed in linearly constrained optimization is based upon the QR factorization

$$N_q = Q\begin{bmatrix} R \\ 0 \end{bmatrix} = [Q_1 \quad Q_2]\begin{bmatrix} R \\ 0 \end{bmatrix}$$

and the Cholesky factorization of the projected (approximate) Hessian

$$Q_2^T G Q_2 = LL^T$$

This approach is described in detail in [2] and [3]. It is important to note that our multiple drop strategies require the second-order Lagrange multiplier estimates

$$\alpha^q = (N_q^T N_q)^{-1} N^T (g + Gp) = R^{-1} Q_1^T (g + Gp)$$

where

$$p = -Q_2 L^{-T} L^{-1} Q_2^T g,$$

and not the first order estimates

$$\lambda^q = (N_q^T N_q)^{-1} N_q^T g = R^{-1} Q_1^T g,$$

which are usually computed when using this approach.

5. USE IN PRIMAL QUADRATIC PROGRAMMING ALGORITHMS.

We have found the ability to drop more than one constraint at a time from the active set to be very useful in solving positive definite quadratic programs by primal methods. In particular, implementing strategy 1 within a standard primal active set quadratic programming algorithm requires only an additional n-vector of storage and very little change in coding and computation per step.

We present such an algorithm below as algorithm 1. Only a single constraint is dropped at a time from the active set, and this is done only at a point which is a constrained minimizer of the equality constrained quadratic program corresponding to the current active set. To illustrate the ease and efficiency with which strategy 1 can be implemented within algorithm 1, all of the changes to it that are required are underlined. We refer to the resulting algorithm as algorithm 2. Both of these algorithms are described in detail in [6]. The vector u and the sets V and C are what we previously referred to as α^q, J^q and J_2^q. D is the set of constraints dropped since a constrained minimizer was last computed.

Algorithm 1: (Algorithm 2)

0) Initialization

Find a feasible constrained minimizer x. Set A to the active set, and compute the primal slacks s(x), the matrices J and R and the dual variables $u = R^{-1}J_1^T g(x)$.

1) Check for optimality:

If $V = \{j \in A \mid u_j < 0\} = \emptyset$, STOP; x is optimal.

Otherwise set $C \leftarrow A$, $D \leftarrow \emptyset$, $d \leftarrow 0$ and $r \leftarrow 0$.

2) Choose constraints to drop:

a) Repeat until $V = \emptyset$.

Choose $p \in V \backslash D$ to drop from the active set.

Set $A \leftarrow A \backslash \{p\}$, $D \leftarrow D \cup \{p\}$, $C \leftarrow C \backslash \{p\}$, $d \leftarrow \underline{d} + u_p v$ where v is the column of $\begin{bmatrix} R \\ 0 \end{bmatrix}$ corresponding to n_p, and update J, R, and d.

b) Compute the step direction in dual space $\bar{r} = R^{-1}d_1$, if k > 0, and set $C \leftarrow \{j \in C \mid \bar{r}_j < r_j\}$, $r \leftarrow \bar{r}$ and $V \leftarrow \{j \in C \mid u_j + r_j < 0\}$.

3) Compute step direction and length:

a) Compute step direction in primal space.

Compute $z = -J_2 d_2$, if k < n, (else set z=0).

b) Compute step length.

(i) Maximum step without violating primal feasibility:

If $v_j \equiv n^T z \geq 0$ for all $j \in K \backslash A$, set $t_1 \leftarrow \infty$; otherwise set $t_1 \leftarrow \min\{-s_j/v_j \mid v_j < 0,\ j \in K \backslash A\} = -s_q/v_q$

(ii) Actual step length:

Set $t \leftarrow \min\{t_1, 1\}$

(iii) Take step in primal and dual spaces.

Set $x \leftarrow x + tz$

$u \leftarrow u + tr$

and $s_j \leftarrow s_j + tv_j$, for all $j \in K \backslash A$.

(Note: z may equal 0 even if $r \neq 0$.)

If t=1 (full step) to to (1).

If t < 1 (partial step) add constraint q to active set as the last constraint; i.e., set $A \leftarrow A \cup \{q\}$, update J, R and d and set $u \leftarrow \binom{u}{0}$, $d \leftarrow (1-t)d$, $r \leftarrow R^{-1}d_1$, $C \leftarrow A$, and $V \leftarrow \{j \in A \mid u_j + r_j < 0\}$ and go to (2a (b)).

6. STEEPEST FACE CRITERIA.

In most active set primal algorithms the usual rule for selecting a constraint to be dropped from amongst a set of candidates (e.g., the set V in the algorithms of the previous section) is to choose the one with the most negative Lagrange multiplier. This is the policy that is followed by most implementations of the simplex method for linear programming. As in the case of linear programming, (e.g., see [12] and [14]) this is not always the best choice.

In [8], it was shown how to choose the constraint to drop in the simplex method so that the resulting descent direction would be along the steepest edge emanating from the current vertex (active set). In the methods considered here, the descent direction followed after a constraint is dropped, is the projected Newton direction $p = -H_{q-1}g$, based upon the metric G. Consequently, it makes sense to choose that constraint to drop which results in a direction $p = -H_{q-1}g$ which is steepest in the G-norm. That is, we want to make this choice so as to minimize

$$\frac{p^Tg}{\|p\|_G} = \frac{p^Tg}{(p^TGp)^{1/2}} = \frac{-g^TH_{q-1}g}{(g^TH_{q-1}GH_{q-1}g)^{1/2}} =$$

$$\frac{-g^TH_{q-1}g}{(g^TH_{q-1}g)^{1/2}} = -(g^TH_{q-1}g)^{1/2}$$

It follows from (14) and the definition of α^q that

$$g^TH_{q-1}g = g^TH_qg + (\alpha_j^q)^2/n_j^{*T}Gn_j^*$$

if the j-th constraint is deleted from the active set.

Therefore, we obtain the following "steepest-face" dropping rule: choose that constraint $\hat{j}$ for which

$$(\alpha_{\hat{j}}^q)^2/\gamma_{\hat{j}}^q = \max\{(\alpha_j^q)^2/\gamma_j^q \mid j \in V\}$$

where

$$\gamma_j^q = n_j^{*T}Gn_j^*$$

and V is the set of constraints which are candidates for dropping (e.g., those with negative Lagrange multipliers).

This rule can also be interpreted as specifying that constraint in the set V to drop for which the resulting equality constrained quadratic program corresponding to the quadratic model has a minimizer of lowest function value. This follows from the fact that

$$Q(p) = 1/2p^TGp + g^Tp = 1/2g^TH_{q-1}G\,H_{q-1}g - g^TH_{q-1}g = 1/2g^TH_{q-1}g \quad .$$

Steepest-face strategies have also been suggested for use in non-simplex active set algorithms for linear programming by Gould [10].

Implementing a steepest-face strategy is more expensive than choosing a constraint from the candidate set V on the basis of the magnitude of the Lagrange multipliers because the quantities γ_j^q are needed for all $j\varepsilon V$. Notice that these quantities are also needed by the recurrence formulas (12) for the b_j^q. How much extra work is required depends very heavily upon the implementation that is used.

This additional amount of work is quite small in the quadratic programming algorithms described in the previous section if one uses the recurrence formulas below for updating the γ_j^q when the active set is changed.

<u>Recurrence formulas for γ_j^q:</u>

Add constraint q+1:

$$\gamma_{q+1}^{q+1} = 1/n_{q+1}^TH_qn_{q+1} = 1/h^Th \;, \qquad h=J^Tn_{q+1}$$

$$\gamma_j^{q+1} = \gamma_j^q + \gamma_{q+1}^{q+1}r_j^2 \;, \qquad r = N_q^*n_{q+1}$$

Drop constraint $\hat{j}$: (first cyclically permute the last $q - \hat{j} + 1$ active set indices to make $\hat{j}$ the last (q-th) one).

$$\gamma_j^{q-1} = \gamma_j^q - u_j^2/\gamma_q^q \;, \qquad u = N_q^*Gn_q^*$$

Since both h and r are computed by the algorithms whether or not a steepest-face dropping rule is used, the updating formulas for adding a constraint require only O(q) additional work. To compute u when a constraint is dropped one cyclically permutes the columns of R as indicated above and applies Givens rotations to transform the resulting upper-Hessenberg matrix to an upper triangular matrix:

$$\begin{bmatrix} \bar{R} & w \\ 0 & \omega \end{bmatrix} .$$

This does not require any additional work since this is already done in the algorithms. The vector $\binom{w}{\omega}$ is what the first q elements of v become after it is updated in step 2 (see [6] for details). One must then solve

$$\bar{R}u = (-1/\omega^2)w$$

for u. This requires $O(q^2)$ work. Also it is easy to show that $\gamma_q^q = 1/\omega^2$.

Since $N_q^* G N_q^{*T} = R^{-1}R^{-T}$ one can also compute any single γ_j^q directly using at most $O(q^2)$ operations.

REFERENCES:

1. FLETCHER, R. A general quadratic programming algorithm, J. Inst. Math. Appl. 7 (1971) 76-91.

2. GILL, P. E. and W. MURRAY Newton-type methods for unconstrained and linearly constrained optimization, Math. Prog. 28 (1974) 311-350.

3. GILL, P. E. and W. MURRAY Numerically stable methods for quadratic programming, Math Prog. 14 (1978) 349-372.

4. GOLDFARB, D. Extension of Newton's method and simplex methods for solving quadratic programs, Numerical Methods for Nonlinear Optimization, Academic Press, London, 1972, 239-254.

5. GOLDFARB, D. Matrix factorizations in optimization of nonlinear functions subject to linear constraints, Math. Prog. 10 (1975) 1-31.

6. GOLDFARB, D. Efficient primal algorithms for strictly convex quadratic programs, Proceedings of the Fourth IIMAS Workshop on Numerical Analysis, Guanajuato, Mexico, July 1984 (ed. J. P. Hennart), "Lecture Notes in Mathematics," Springer-Verlag, 1985.

7. GOLDFARB, D. and A. IDNANI A numerically stable dual method for solving strictly convex quadratic programs, Math. Programming 27 (1983) 1-33.

8. GOLDFARB, D. and J. K. REID A practicable steepest edge simplex algorithm, Math. Programming 12 (1977) 361-373.

9. GOLUB, G. H. and M. A. SAUNDERS Linear least squares and quadratic programming, Integer and Nonlinear Programming (ed. J. Abadie) North Holland Publ. Co., 1970, 229-256.

10. GOULD, N. I. M. The Generalized Steepest-Edge for Linear Programming, Part I: Theory, Dept. of Combinatorics and Optimization Tech. Rept., Univ. of Waterloo, Ontario, 1984.

11. IDNANI, A. U. Numerically stable dual projection methods for solving positive definite quadratic programs, (Ph.D. thesis) City College of New York, Dept. of Computer Science, New York, 1980.

12. KUHN, H. W. and R. E. QUANDT An experimental study of the simplex method, Proc. of Symposia in Applied Maths Vol. XV (ed. Metropolis et al.), A. M. S., 1963.

13. ROSEN, J. B. The gradient projection method for nonlinear programming, Part 1 - linear constraints, SIAM Journal 8 (1960) 181-217.

14. WOLFE, P. and J. CUTLER Experiments in linear programming, Recent Advances in Mathematical Programming (ed. Graves and Wolfe), McGraw-Hill, 1963.

APPENDIX:

We give an example below which illustrates that:

(i) strategy 2 can allow one to drop more constraints than strategy 1,

(ii) strategy 2 does not necessarily determine the "best" feasible direction, and

(iii) $\alpha^q_q < 0$ is not sufficient to ensure the feasibility of the direction $-H_{q-1}g$ if constraints have already been dropped at the current point.

Consider the quadratic program

$$\text{minimize} \quad q(x) = 1/2(x_1^2+x_2^2+x_3^2)$$

$$\text{subject to} \quad \begin{aligned} x_1 - 4x_2 &\geq -4 \\ -x_1 - 4x_2 &\geq -4 \\ 4x_2 - 1/4x_3 &\geq 3 \end{aligned}$$

At $x = (0,1,4)^T$ all constraints are tight (i.e., $E = \{1,2,3\}$).

Let the active set $A_3 = E$. Hence,

$$N_3 = \begin{bmatrix} 1 & -1 & 0 \\ -4 & -4 & 4 \\ 0 & 0 & -1/4 \end{bmatrix}, \quad N_3^* = \begin{bmatrix} 1/2 & -1/8 & -2 \\ -1/2 & -1/8 & -2 \\ 0 & 0 & -4 \end{bmatrix}, \quad \alpha^3 = \begin{bmatrix} -8\ 1/8 \\ -8\ 1/8 \\ -16 \end{bmatrix},$$

and $J^3 = J^3_1 = \{1,2,3\}$.

Choosing constraint 3 to drop we obtain $A_2 = \{1,2\}$,

$$N_2 = \begin{bmatrix} 1 & -1 \\ -4 & -4 \\ 0 & 0 \end{bmatrix}, \quad N_2^* = \begin{bmatrix} 1/2 & -1/8 & 0 \\ -1/2 & -1/8 & 0 \end{bmatrix}, \quad \alpha^2 = \begin{bmatrix} -1/8 \\ -1/8 \end{bmatrix},$$

$$-H_2 g = (0,0,-4)^T, \quad J^2_1 = \{1,2\}, \text{ and } J^2_2 = J^2 = \emptyset .$$

At this point strategy 1 will not allow any more constraints to be dropped.

However, since $J_1^2 \neq \emptyset$, if strategy 2 is used we can <u>try</u> to drop either constraint 1 or 2. Choosing constraint 2, we compute

$$b_2^1 = -8/17 \quad \text{and} \quad b_3^1 = -13/17$$

using the recurrence formulas for b_j^q and c_j^q; consequently, we can drop this constraint and obtain $A_1 = \{1\}$,

$$N_1 = \begin{bmatrix} 1 \\ -4 \\ 0 \end{bmatrix}, \quad N_1^* = [1/17, -4/17, 0], \quad \alpha^1 = -4/17,$$

$$-H_1 g = (-4/17, -1/17, -4), \quad J_1^1 = \{1\}, \text{ and } J_2^1 = J^1 = \emptyset.$$

Now if we apply strategy 2, and try to drop constraint 1, we find that

$$b_1^0 = -4, \quad b_2^0 = -4, \text{ and } b_3^0 = 3.$$

Thus we cannot drop this constraint since doing so and moving in the direction $-H_0 g = -g = (0,-1,-4)^T$ will violate constraint 3 which was previously dropped. Note that $\alpha_1^1 = -4/17 < 0$.

Furthermore, it is easy to verify that solving the appropriate least distance programming problem at the point (0,1,4) yields $p = (0,-65/257,-1048/257)^T$ as the best feasible direction yielding the point (0,192/257,-12/257), lying only on constraint 3, as the solution to our quadratic program; i.e., we should have dropped constraints 1 and 2 from the active set rather than 3 and 2.

Even if we had initially chosen to drop constraints 1 or 2 from the active set, we would not necessarily have obtained the "optimal" active set and, hence, the "best" feasible direction. As before strategy 1 would not have allowed more than one constraint to be dropped, while strategy 2 would have resulted in two constraints being dropped. However, which two would have depended upon the order in which the constraints in J_1^2 were considered.

D. Goldfarb
Department of Industrial Engineering
and Operations Research
Columbia University
Seeley W. Mudd Building
New York
NY 10027
U.S.A.

S-P HAN

Optimization by updated conjugate subspaces

1. INTRODUCTION

We consider in this paper the unconstrained minimization problem

$$\min_{R^n} f(x) \tag{1.1}$$

where $f:R^n \to R$ is twice continuously differentiable. For such a problem the quasi-Newton methods and the conjugate direction methods are deemed to be very effective. Here we propose a class of methods which possess attractive features of those methods and are also suitable for parallel computation.

The quasi-Newton method is based on quadratic approximation. Having an estimate x of a solution of (1.1) and an estimate B of the Hessian $\nabla^2 f(x)$, we make use of the approximation

$$f(x+d) \approx q(d) := f(x) + \nabla f(x)^T d + \tfrac{1}{2}d^T Bd \tag{1.2}$$

and find a search direction d by solving

$$\min_{R^n} q(d) ,$$

which is generally done by solving the linear system of equations

$$Bd = -\nabla f(x) . \tag{1.3}$$

The matrix B is usually required to be positive definite to ensure a descent direction for f. After a line search for determining a stepsize λ, a new estimate $\bar{x}$ is given by $\bar{x} = x + \lambda d$. If this new estimate is still not satisfactory and further improvement is needed, we repeat this calculation with x replaced by $\bar{x}$ and B replaced by a new matrix $\bar{B}$ which satisfies the quasi-Newton equation

$$\bar{B}s = y \tag{1.4}$$

where $s = \bar{x} - x$ and $y = \nabla f(\bar{x}) - \nabla f(x)$. The reasons for the quasi-Newton equation to hold are many and well stated in the literature [see, for example, 1,2,4]. Among many schemes for generating $\bar{B}$ to satisfy (1.4) the most successful one is the BFGS formula:

$$\bar{B} = B + \frac{yy^T}{y^Ts} - \frac{Bss^TB}{s^TBs} . \qquad (1.5)$$

The efficiency of the method lies on the well-known fact that Hessians of f can be very well approximated by such updated matrices which depend only on the first order derivatives.

Another class of methods which are relevant to our approach are the conjugate direction methods. These methods can also be viewed as based on the quadratic approximation (1.2) with a positive definite B. But, instead of solving (1.3) directly, we search along directions $z_1,\ldots z_n$ which satisfy the conjugacy condition

$$z_i^TBz_j = 0 \qquad \text{if} \quad i \neq j .$$

Such search is usually done iteratively as

$$d_{k+1} = d_k + \lambda_k z_k$$

with d_1 an arbitrary vector and λ_k stepsizes. It is well known that if exact line searches are carried out with respect to the quadratic function q, then the solution d in (1.3) can be found in at most n iterations. There are various ways to extend this technique to the general nonlinear problem (1.1) [see, for example, 3,5,6,7,8]; and in the special case of a conjugate gradient method the vectors z_i are usually derived from gradients of f.

Though effective as they are for solving the unconstrained problem (1.1), the usual implementations of the quasi-Newton methods and the conjugate direction methods are not appropriate for parallel computation, for which the following straightforward pattern search method seems suitable. At a point x we consider subspaces $T_1,\ldots T_m$ which satisfy

(a) $R^n = T_1 + \ldots + T_m$;

(b) $\{T_1, \ldots T_m\}$ are linearly independent; that is,

$$x_1 + \ldots + x_m = 0, \quad x_i \in T_i \Rightarrow \quad x_1 = \ldots = x_m = 0 .$$

Then for each i, (i = 1,...,m), we find a vector d_i by solving

$$\begin{array}{ll} \min & f(x+d) \\ \text{s.t.} & d \in T_i \end{array} \qquad (1.6)$$

or

$$\begin{array}{ll} \min & q(d) \\ \text{s.t.} & d \in T_i \end{array} \qquad (1.7)$$

where q is defined as in (1.2). Then the search direction is given by

$$d = d_1 + \ldots + d_m . \qquad (1.8)$$

Here the key feature is that the computation of d_i in (1.6) or (1.7) can be done parallelly.

It is observed that the ideas behind the quasi-Newton methods, the conjugate direction methods and the aforementioned pattern search method are not inconsistent. It is feasible to develop a method which possesses all three main respective features of the three methods. A main theme of the paper is to discuss how we can design a method in which a quasi-Newton scheme is used to provide estimates of Hessians, search directions or subspaces are chosen to be conjugate with respect to those matrices, and searches in these directions or subspace are carried out parallelly.

2. QUASI-NEWTON METHODS VIA CONJUGATE SUBSPACES

We recall below the definition of conjugate subspaces and a fundamental theorem on which our approach is based.

Definition 2.1. A subspace T in R^n is said to be the <u>conjugate sum</u> of subspaces $T_1, \ldots, T_m$ with respect to a symmetric matrix B which is positive definite on T if

(a) $T = T_1 + \ldots + T_m$;

(b) for any $i \neq j$, $x_i \in T_i$ and $x_j \in T_j$

$$x_i^T B x_j = 0 .$$

In this case, T is written as

$$T = T_1 (+)_B \ldots (+)_B T_m .$$

Theorem 2.2 Let B be symmetric and positive definite on a subspace T and let $T = T_1 (+)_B \ldots (+)_B T_m$. If d_i is the minimum point of $q(d) := \frac{1}{2} d^T B d + c^T d$

in T_i then the vector $\bar{d} := d_1 + \ldots + d_m$ is the minimum point of q in T.

Proof: Because d_i is the minimum point of q in T_i we have that

$$\nabla q(d_i) \in T_i^{\perp} .$$

On the other hand, the conjugacy of T_i's implies that for $j \neq i$,

$$Bd_j \in T_i^{\perp} .$$

Therefore, we have that

$$\begin{aligned} \nabla q(\bar{d}) &= Bd_i + c + \sum_{\substack{j=i \\ j\neq i}}^{m} Bd_j \\ &\in T_i^{\perp} . \end{aligned}$$

Consequently.

$$\begin{aligned} \nabla q(\bar{d}) \in \bigcap_{i=1}^{m} T_i^{\perp} &= (T_1 + \ldots + T_m)^{\perp} \\ &= T^{\perp} . \end{aligned}$$

Q.E.D.

If the columns of a matrix Z_i form the basis of the subspace T_i, then the vectors d_i and $\bar{d}$ in theorem 2.2 can be computed as

$$d_i = - Z_i(Z_i^T BZ_i)^{-1} Z_i^T c$$

$$\bar{d} = - \left(\sum_{i=1}^{m} Z_i(Z_i^T BZ_i)^{-1} Z_i^T \right) c .$$

In the special case that $T_i = \text{span}(z_i)$ is one dimensional, then we have the familiar formula

$$\bar{d} = - \sum_{i=1}^{m} \frac{z_i^T c}{z_i^T Bz_i} z_i .$$

Because we will restrict ourselves to unconstrained problems, we only consider the case $T = R^n$. In this case $\bar{d} = - B^{-1} c$ and, consequently, we have the following corollary.

Corollary 2.3: If $\nabla^2 f(x)$ is positive definite and the search subspaces $T_1, \ldots, T_m$ are conjugate with respect to $\nabla^2 f(x)$, then the search direction d generated by the straightforward pattern search method using (1.7) and (1.8) is the Newton direction

$$d = -\nabla^2 f(x)^{-1} \nabla f(x) .$$

The above corollary indicates that the straightforward pattern search method can be very efficient if the matrix B in (1.7) is close to the Hessian $\nabla^2 f(x)$ and the search subspaces $T_1, \ldots, T_m$ are conjugate with respect to B. Because some quasi-Newton schemes can generate very good approximations to Hessians, we are motivated by this result to study how to update subspaces which are conjugate with respect to such approximations.

Suppose at a certain iteration we have a positive definite symmetric matrix B and subspaces $T_1, \ldots, T_m$ such that

$$R^n = T_1 (+)_B \cdots (+)_B T_m . \tag{2.1}$$

After a new estimate $\bar{x}$ is computed by (1.7), (1.8) and a line search, and after a new matrix $\bar{B}$ is updated, we want to continue the same process and to get subspaces $\bar{T}_1, \ldots, \bar{T}_m$ such that

$$R^n = \bar{T}_1 (+)_{\bar{B}} \cdots (+)_{\bar{B}} \bar{T}_m . \tag{2.2}$$

We are interested in knowing whether such subspaces $\bar{T}_1, \ldots, \bar{T}_m$ can be updated easily from subspaces $T_1, \ldots, T_m$. This certainly depends on the corresponding scheme for updating the matrix $\bar{B}$. For the case of the BFGS formula (1.5) this can be easily done as described and justified in the following theorem.

Theorem 2.4: Let B be positive definite and symmetric and s,y be two vectors such that $s^T y > 0$ and let $T, T_1, \ldots, T_m$ be subspaces such that

$$T = T_1 (+)_B \cdots (+)_B T_m .$$

If

$$A := I - \frac{sy^T}{y^T s} + (s^T y \; s^T Bs)^{-\frac{1}{2}} \; ss^T B \tag{2.3}$$

$$\bar{T}_i := AT_i := \{u : u = Av, v \in T_i\} \tag{2.4}$$

then

$$AT = \bar{T}_1 \oplus_{\bar{B}} \dots \oplus_{\bar{B}} \bar{T}_m$$

where $\bar{B}$ is obtained from B, s and y from the BFGS formula (1.5). If, in addition, $T = R^n$ then $AT = R^n$.

<u>Proof</u>: This follows directly from the equation $A^T\bar{B}A = B$. The last statement follows from the nonsingularity of A because

$$\det(A)^2 \det(\bar{B}) = \det(B) \neq 0 . \qquad \text{Q.E.D.}$$

In view of theorem 2.4 we can carry out one iteration of the BFGS method through the conjugate space approach as follows. Having x, B and subspaces $T_1, \dots, T_m$ satisfying (2.1), we solve for d_i, $(i = 1, \dots, m)$,

$$\begin{aligned} \min \quad & q(d) := f(x) + \nabla f(x)^T d + \tfrac{1}{2} d^T B d \\ \text{s.t.} \quad & d \in T_i \end{aligned} \tag{2.5}$$

As mentioned before, this can be done by choosing a matrix Z_i whose columns form a basis of T_i and by solving for w_i

$$\min_w f(x) + \nabla f(x)^T Z_i w + \tfrac{1}{2} w^T Z_i^T B Z_i w \tag{2.6}$$

and setting $d_i = Z_i w_i$. Then a search direction d is computed by $d = d_1 + \dots + d_m$ and a new estimate $\bar{x}$ is computed by $\bar{x} = x + \lambda d$ with a stepsize λ. To continue we update subspaces $\bar{T}_1, \dots, \bar{T}_m$ by (2.3) and (2.4) and update $\bar{B}$ by (1.5).

Initially, from a theoretical viewpoint, we can start from any set of subspaces as long as they span the whole space R^n and are linearly independent. This is because if Z_i is a matrix whose columns form a basis of T_i and Z is the n×n matrix $[Z_1, \dots, Z_m]$ then the subspaces $T_1, \dots, T_m$ are conjugate with respect to the matrix $(ZZ^T)^{-1}$, which is considered as an initial approximation to a Hessian of f. Therefore, ideal starting subspaces should be those spanned by columns of a matrix Z which satisfies

$$ZZ^T \approx \nabla^2 f(x)^{-1}$$

or the approximate conjugacy condition

$$Z^T \nabla^2 f(x) Z \approx I .$$

Since such a matrix is not easily available, we may simply choose $Z = I$ and the starting subspaces are generated by the coordinate unit vectors. This is equivalent to the choice of $B^{(o)} = I$ in a usual implementation of the BFGS method, as is often done in practice.

Throughout the calculation, the subspaces $T_1, \ldots, T_m$ can always be represented by an $n \times n$ nonsingular matrix Z partitioned into column blocks $[Z_1, \ldots, Z_m]$ with each block forming a basis of the corresponding subspace. In so doing the subspace updating can be easily carried out by premultiplying Z by the updating matrix A given in (2.3) to form bases of the new subspaces $\bar{T}_1, \ldots, \bar{T}_m$ as follows:

$$
\begin{aligned}
\bar{Z} \;:=\;& [\bar{Z}_1, \ldots, \bar{Z}_m] \\
=\;& [AZ_1, \ldots, AZ_m] \\
=\;& AZ \;.
\end{aligned}
$$

The special basis Z_i of T_i obtained through updating from the initial $Z_i^{(o)}$ will be called the <u>inherent basis</u> of T_i. Clearly the inherent basis $Z_i^{(k)}$ at the kth iteration can be expressed as

$$Z_i^{(k)} = A^{(k)} \ldots A^{(1)} Z_i^{(o)}$$

where $A^{(k)}$ is the updating matrix at the kth iteration.

It is noted that the vector d_i in (2.5) is independent of a choice of a basis for T_i. However, it is of great advantage to use inherent bases in calculating w_i in (2.6), because such bases carry information about the matrix B and can help us to avoid computing it. This is a consequence of the following theorem.

Theorem 2.5: If the column blocks of $Z^{(k)} = [Z_1^{(k)}, \ldots, Z_m^{(k)}]$ are the inherent bases of $T_1^{(k)}, \ldots, T_m^{(k)}$ respectively then

$$Z^{(k)T} B^{(k)} Z^{(k)} = I \;.$$

<u>Proof</u>: As indicated in theorem 2.4 the updating matrix $A^{(k)}$ satisfies

$$A^{(k)T} B^{(k)} A^{(k)} = B^{(k-1)} \;.$$

Then it follows from $B^{(o)} = (Z^{(o)} Z^{(o)T})^{-1}$ that

$$Z^{(k)T} B^{(k)} Z^{(k)} = Z^{(k-1)T} A^{(k)T} B^{(k)} A^{(k)} Z^{(k-1)}$$

$$= Z^{(k-1)T} B^{(k-1)} Z^{(k-1)}$$

$$\vdots$$

$$= Z^{(o)T} B^{(o)} Z^{(o)}$$

$$= I .$$ Q.E.D.

Therefore, by $Z_i^T B Z_i = I$ the subproblem (2.6) becomes

$$\min_w f(x) + \nabla f(x)^T Z_i w + \tfrac{1}{2} w^T w$$

and w_i can be computed without knowing the matrix B as follows:

$$w_i = - Z_i^T \nabla f(x) .$$

The matrix B appearing in the updating matrix A can also be removed by observing that A contains only Bs and Bs = $-\lambda g$ where $g = \nabla f(x)$. Hence, the matrix A can now be rewritten without B as

$$A = I - \frac{sy^T}{y^T s} - \left(\frac{-\lambda}{g^T s y^T s} \right)^{\frac{1}{2}} sg^T . \tag{2.7}$$

Now we can state how to implement the BFGS method in a parallel computer as below.

Implementation of the BFGS method in parallel computation:

Start from a point $x^{(o)}$ and an $n \times n$ nonsingular matrix $Z^{(o)} = [Z_1^{(o)}, \ldots, Z_m^{(o)}]$ partitioned into column blocks.

At the k-th iteration, having x and $Z = [Z_1, \ldots, Z_m]$, compute $\bar{x}$ and update $\bar{Z} = [\bar{Z}_1, \ldots, \bar{Z}_m]$ as follows:

(1) For i = 1, ..., m solve for w_i and d_i by

$$w_i = - Z_i^T \nabla f(x)$$

$$d_i = Z_i w_i$$

(2) Find a search direction d by

$$d = d_1 + \ldots + d_m .$$

(3) Determine a stepsize λ and set $\bar{x} = x + \lambda d$.

(4) Check convergence; if not, set A as in (2.7) and

$$\bar{Z} = AZ.$$

It is noted that the matrix A is of the form

$$A = I - su^T$$

where

$$u := \left(\frac{1}{y^T s}\right) y + \left(\frac{-\lambda}{s^T g s^T y}\right)^{\frac{1}{2}} g.$$

Once u is calculated, the computation of $\bar{z} = Az$ can be easily done as $\bar{z} = Az = z - (u^T z)s$. Therefore, the calculation of $\bar{Z}$ is not a very costly process.

As for the other quasi-Newton methods, they can also be implemented in a similar manner with only the updating matrix A different. For the DFP case the matrix A is given by

$$A_{DFP} = I - \frac{Hyy^T}{y^T Hy} + \frac{sy^T}{(y^T s \; y^T Hy)^{\frac{1}{2}}}$$

where $H = B^{-1}$. We may also consider the following family of methods with A being a convex combination of A_{BFGS} and A_{DFP}:

$$A_\theta = \theta A_{BFGS} + (1-\theta) A_{DFP} \qquad \theta \in [0,1] .$$

This generally does not satisfy the quasi-Newton equation and is not Broyden's family, for which the corresponding matrix A is even more complicated. But, in any case, besides the BFGS method, the other quasi-Newton methods seem not suitable for the conjugate space approach, not only because they are inferior in efficiency to the BFGS method but also because it is less convenient to compute A.

3. THE UPDATED CONJUGATE SUBSPACE METHOD

The approach described in the previous section provides some flexibilities in implementing a quasi-Newton method. We study in this section how to exploit such flexibilities in order to improve the performance of the method. For doing so we consider the following general method based on the concept of updating conjugate subspaces.

The updated conjugate subspace method

Start from any point $x^{(o)}$ and any nxn nonsingular matrix $Z^{(o)} = [Z_1^{(o)}, \ldots, Z_m^{(o)}]$ partitioned into column blocks.

At the k-th iteration, having x and $Z = [Z_1, \ldots, Z_m]$ we compute a new estimate $\bar{x}$ and update a matrix $\bar{Z} = [\bar{Z}_1, \ldots, \bar{Z}_m]$ as follows:

(1) for $i = 1, \ldots, m$, compute w_i which solves

$$\min f(x) + \nabla f(x)^T Z_i w + \tfrac{1}{2} w^T M_i w \qquad (3.1)$$

where M_i are positive definite symmetric matrices.

(2) For $i = 1, \ldots, m$, set $d_i = Z_i w_i$.

(3) Set the search direction $d = d_1 + \ldots + d_m$.

(4) Determine a stepsize λ and set $\bar{x} = x + \lambda d$.

(5) Check convergence; if not, update $\bar{Z}$ by $\bar{Z} = AZ$, where A is an $n \times n$ nonsingular matrix.

A special method is determined once the matrices A and M_i are specified. How to choose such matrices is a main concern of our following discussion. The theorem below gives us some clues for making such choices.

Theorem 3.1: Let $\nabla^2 f(x)$ be positive definite. If the subspaces $T_i := \text{range}(Z_i)$ are conjugate with respect to $\nabla^2 f(x)$ and if for $i=1,\ldots,m$, $M_i = Z_i^T \nabla^2 f(x) Z_i$, then the search direction d is the Newton direction

$$d = -\nabla^2 f(x)^{-1} \nabla f(x) .$$

<u>Proof</u>: This follows directly from

$$\nabla^2 f(x)^{-1} = \sum_{i=1}^{m} Z_i (Z_i^T \nabla^2 f(x) Z_i)^{-1} Z_i^T$$

$$d = -\left(\sum_{i=1}^{m} Z_i M_i^{-1} Z_i^T \right) \nabla f(x) \qquad \text{Q.E.D.}$$

Therefore, for the method to produce a direction close to a Newton direction we should establish to a high degree

(a) conjugacy of $T_1, \ldots, T_m$ with respect to $\nabla^2 f(x)$;

(b) $M_i \approx Z_i^T \nabla^2 f(x) Z_i$, $i = 1, \ldots, m$.

The first requirement is attempted to be fulfilled by an appropriate choice of the updating matrix A; while the second by a choice of M_i.

3.1 <u>Strategies for choosing A</u>

(1) $A = I$

This is a commonly known pattern search method in which the search subspaces remain fixed throughout calculation.

(2) A is orthogonal

In this case, the search subspaces are allowed to rotate after each iteration.

(3) A is chosen by a quasi-Newton scheme

This provides an implementation of a quasi-Newton method in parallel computation.

The choice of A given in (2.7) is certainly preferred for the obvious reasons that it is corresponding to the successful BFGS method and also it is simple. Therefore, we always assume that A is chosen by (2.7) in the following discussion.

3.2 Strategies for choosing M_i

(1) $M_i = I$

This resuls in the BFGS method, though the implementation is different from the usual one. The choice of $M_i = I$ can be viewed as based on the approximation

$$Z_i^T \nabla^2 f(x) Z_i \approx I$$

and hence based on the assumption that the basis vectors of the subspace T_i = range (Z_i) in Z_i are very conjugate with respect to $\nabla^2 f(x)$ and also well scaled. When the inherent bases are used, $M_i = I$ is a very sensible choice because the basis vectors in Z_i are just as conjugate with respect to $\nabla^2 f(x)$ as the subspaces $T_1, \ldots, T_m$ are, for they are updated in the same manner.

(2) $M_i = \text{diag}\{\beta_1, \ldots, \beta_r\}, \quad \beta_j > 0$

This becomes a scaled BFGS method with scaling factors β_j. Because we want $M_i \approx Z_i^T \nabla^2 f(x) Z_i$, a possible choice for β_j is to let β_j be an estimate of the j-th diagonal element of $Z_i^T \nabla^2 f(x) Z_i$. When $Z_i = [z_1, \ldots, z_r]$ we can choose

$$\beta_j = \frac{1}{\sigma_j} z_j^T (\nabla f(x+\sigma_j z_j) - \nabla f(x))$$

provided the right-hand side quantity is positive, where $\beta_j > 0$ is an increment size. This certainly needs more gradient evaluations.

(3) Exact minimization over subspaces

To calculate the vector d_i in the method we may minimize the function f itself in the subspace T_i to solve the problem

$$\min_{\text{s.t. } d \in T_i} f(x + d) . \tag{3.2}$$

This can also be done by solving for w_i

$$\min_w f(x + Z_i w) \tag{3.3}$$

and set $d_i = Z_i w_i$. In this case the matrix M_i is not given explicitly in advance. But, it can be determined after d_i is computed, though it never needs to be known in practice. Let

$$G_i := \int_0^1 \nabla^2 f(x + td_i)dt .$$

Because d_i solves (3.2) we have that $Z_i^T \nabla f(x + d_i) = 0$ which implies

$$\begin{aligned} Z_i^T G_i Z_i w_i &= Z_i^T G_i d_i \\ &= Z_i^T (\nabla f(x + d_i) - \nabla f(x)) \\ &= - Z_i^T \nabla f(x) . \end{aligned}$$

Therefore, we have that

$$w_i = - (Z_i^T G_i Z_i)^{-1} Z_i^T \nabla f(x) .$$

When f is strictly convex, it may be considered as the case that the matrix M_i in the method is chosen to be $M_i = Z_i^T G_i Z_i$, which is clearly a good approximation to $Z_i^T \nabla^2 f(x) Z_i$.

(4) A bi-level quasi-Newton method

As in (3) we can carry out one or several iterations of a quasi-Newton method for solving the unconstrained minimization problem (3.3). When the dimension of the subspace T_i is small, the subproblem (3.3) can be comparably easy to solve. In this case quasi-Newton schemes are used to update the matrix A in the whole space R^n and also to update the smaller matrices M_i's in the subspaces.

(5) A combined finite-difference Newton and quasi-Newton method

A straightforward way to achieve $M_i \approx Z_i^T \nabla^2 f(x) Z_i$ is by finite differences. More specifically, if $r := \dim(T_i)$ and $Z_i = [z_1, \ldots, z_r]$ we can define the $r \times r$ matrix $M_i = [\alpha_{pq}]$ by

$$\alpha_{pq} := \begin{cases} \frac{1}{\sigma_{pq}} z_p^T(\nabla f(x + \sigma_{pq} z_q) - \nabla f(x)) , & 1 \leq q \leq p \leq r \\ \alpha_{qp} , & 1 \leq p < q \leq r \end{cases}$$

where σ_{pq} are increment sizes. Here, we use the BFGS scheme to establish the conjugacy of the subspaces $T_1, \ldots, T_m$ with respect to $\nabla^2 f(x)$, while we use a finite difference scheme to achieve $M_i \approx Z_i^T \nabla^2 f(x) Z_i$. Consider the following two extreme cases.

(a) $T_i = \text{span}(z_i)$ is one dimensional.

In this case the matrix $M_i = [\alpha_i]$ is just a scalar and is given by

$$\alpha_i := \frac{1}{\sigma_i} z_i^T (\nabla f(x + \sigma_i z_i) - \nabla f(x)).$$

Consequently the search direction d is given by

$$d = -\left(\sum_{i=1}^{n} \frac{1}{\alpha_i} z_i z_i^T\right) \nabla f(x)$$

$$= -ZD^{-1}Z^T \nabla f(x)$$

where $Z := [z_1, \ldots, z_n]$ and $D := \text{diag}\{\alpha_1, \ldots, \alpha_n\}$. Therefore, it becomes a scaled BFGS method as described in (2).

(b) $T_1 = R^n$

In this case M_1 is an $n \times n$ matrix which is a finite difference approximation to $\nabla^2 f(x)$ with increment vectors $z_1, \ldots, z_n$ instead of the coordinate unit vectors usually used in an ordinary finite difference scheme.

In view of the two extreme cases above, we observe that when the dimensions of the subspaces $T_1, \ldots, T_m$ are small the method behaves more like a scaled BFGS method; while when they are large the method is more like a finite difference Newton method. Recall that to form subspaces is just to group together the basis vectors; the procedure is simple and the dimensions of the subspaces can be easily controlled. Therefore, an adaptive strategy can be designed so that the dimensions of the subspaces are adjusted iteratively so as to favour one method or the other during the process of computation.

We remark that the positive definiteness of the matrix M_i is essential and some procedure should be incorporated into the finite difference scheme to preserve this property. It is also worth mentioning that, though the finite difference calculation is in general very costly, this problem becomes less serious when the dimensions of the subspaces are small and the calculation is done parallelly.

4. A PROCEDURE FOR IMPROVING CONJUGACY

The efficiency of the updated conjugate subspace method depends very much on the conjugacy of the subspaces $T_1, \ldots, T_m$ with respect to the Hessian $\nabla^2 f(x)$. In this section we propose a procedure for improving such conjugacy so as to accelerate convergence. This procedure can be best explained in a simple example. Suppose we are working on a space of dimension 6 and have the current basis vectors $\{z_1, \ldots, z_6\}$ such that $T_1 = \text{span}(z_1, z_2, z_3)$ and $T_2 = \text{span}(z_4, z_5, z_6)$. We can apply a procedure such as the Gram-Schmidt procedure in each subspace to generate vectors that are conjugate with respect to the Hessian $\nabla^2 f(x)$. Specifically, we compute $\{u_1, u_2, u_3\}$ from $\{z_1, z_2, z_3\}$ and $\{u_4, u_5, u_6\}$ from $\{z_4, z_5, z_6\}$ as follows:

$$u_1 = z_1$$

$$u_2 = z_2 - \beta_{21} u_1$$

$$u_3 = z_3 - \beta_{31} u_1 - \beta_{32} u_2$$

$$u_4 = z_4$$

$$u_5 = z_5 - \beta_{54} u_4$$

$$u_6 = z_6 - \beta_{64} u_4 - \beta_{65} u_5$$

where

$$\beta_{ij} := \frac{z_i^T y_j}{u_j^T y_j}, \qquad y_j := \nabla f(x + u_j) - \nabla f(x) .$$

Here, as in the derivation of many well-known conjugate gradient methods, we use the approximation $\nabla^2 f(x) u_j \approx y_j$ to achieve

$$\beta_{ij} \approx \frac{z_i^T \nabla^2 f(x) u_j}{u_j^T \nabla^2 f(x) u_j}$$

without calculating the Hessian $\nabla^2 f(x)$. Then we can regroup the basis vectors $\{u_1, \ldots, u_6\}$ to form new subspaces $\hat{T}_1$ and $\hat{T}_2$ such as $\hat{T}_1 = \text{span}(u_2, u_3, u_4)$ and $\hat{T}_2 = \text{span}(u_5, u_6, u_1)$. It is believed that the conjugacy between u_i and $\{u_2, u_3\}$ can contribute somewhat to the conjugacy of the subspaces $\hat{T}_1$ and $\hat{T}_2$ and, hence, can improve the performance of the method.

5. CONCLUSION

The quasi-Newton method has been very successful for solving smooth unconstrained optimization problems. Due to the advent of parallel computers, implementations of the method suitable for parallel computation become necessary. The approach presented in this paper is an attempt to accomplish this goal. Since the parallel computer is still very much in its infancy, it is too early to pass a judgement on the effectiveness of our approach. However, even without its application to parallel computation, the approach still possesses some merits, mainly because it enables us to decompose a large problem into smaller ones. For example, it can be used to solve large and sparse optimization problems. Furthermore, the flexibilities it provides in implementing a quasi-Newton method also allow us to incorporate into the method some procedures, such as those described in Section 3, to improve its performance.

ACKNOWLEDGEMENTS

The research was supported in part by the NSF under Grant MDS 82-03603 and by the British SERC. Most of the work has been done during the author's visit to the University of Cambridge. The author would like to express his thanks to Prof. M.J.D. Powell for his arrangement of the visit and for many inspiring discussions and valuable suggestions.

REFERENCES

1. DENNIS, J.E. & SCHNABEL, R. Numerical Methods for Unconstrained Optimization and Nonlinear Equations, Prentice Hall, New Jersey, 1983.
2. FLETCHER, R. Practical Methods of Optimization, Vol. 1, John Wiley &

Sons, New York, 1980

3. FLETCHER, R. & REEVES, C.M. Function minimization by conjugate gradients, Computer J. 7 (1964) 149-154.
4. GILL, P.E., MURRAY, W. & WRIGHT, M.H. Practical Optimization, Academic Press, London, 1981.
5. HESTENES, M.R. Conjugate-Direction Methods in Optimization, Springer-Verlag, Berlin, 1980.
6. POLAK, E. & RIBIÉRE, G. Note sur la convergence de méthodes de directions conjugée, Rev. Fr. Inform. Rech. Oper. 16-R1 (1969) 35-43.
7. POLYAK, B.T. The conjugate gradient method in extreme problems, USSR Comp. Maths. and Maths. Phys. 9 No.4 (1969) 94-112.
8. POWELL, M.J.D. An efficient method for finding the minimum of a function of several variables without calculating derivatives, Computer J. 7 (1964) 155-162.

S.-P. Han
Department of Mathematics
University of Illinois
Urbana, Illinois 61801
USA

A ISERLES

Order stars and stability barriers

1. INTRODUCTION

A finite difference scheme for the numerical solution of a differential equation is, essentially, an algebraic structure that imposes a relationship between approximations to the solution (and, possibly, its derivatives) at different grid points. The choice of a "good" scheme entails the reconciliation of competing aims. Obviously, one is interested in good local approximation of the underlying differential equation - this can be expressed by the order of the scheme. However, to infer from local to global behaviour it is necessary to impose stability. Moreover, one might well be interested in the numerical scheme reproducing faithfully certain aspects of the differential equation which are essential on physical grounds, e.g. preservation of positivity. Last - but not least - an account must be taken of the likely numerical cost in solving the (perhaps nonlinear) algebraic equations.

Some of the features of the scheme are qualitative - a scheme is either stable (in a certain sense) or it is not - whilst others are quantitative and a matter of degree. Thus, it makes sense to pose the following problem: subject to stability (or monotonicity or any other qualitative feature), what is the maximal order or what is the form which is most amenable to actual computational work.

The oldest (and most famous) barrier is due to Dahlquist [3]: the maximal order of the linear multistep method

$$\sum_{k=0}^{m} \alpha_k Y_{n+k} = h \sum_{k=0}^{m} \beta_k f(t_{n+k}, Y_{n+k}), \qquad n \geq 0, \tag{1}$$

for the ordinary differential equation

$$y' = f(t,y), \qquad t \geq t_0,$$
$$y(t_0) = y_0, \tag{2}$$

subject to zero-stability, is $2[\frac{1}{2}(m+1)]$. Here $t_k = t_o + kh$ and Y_k is an approximation to $y(t_k)$, $k \geq 0$.

Although several stability barriers were derived by standard methods [2,4,5,23,28], the subject gathered substantial momentum with the introduction of order stars by Wanner, Hairer and Nørsett [31] in 1978. It led to numerous new results for both ordinary and partial differential equations of evolution.

The order-star approach to stability barriers is, essentially, a set pattern of two steps: first, the numerical problem is translated into a problem in complex approximation theory. Next, order stars are used to express the approximation-theoretic problem in combinatorial terms. In Section 2 we follow this path step-by-step, outlining the proof of an order barrier for explicit semi-discretised approximations to the first order linear hyperbolic advection equation $\partial u/\partial t = \partial u/\partial x$. Sections 3 and 4 are devoted to a review of existing barriers for ordinary and partial differential equations, respectively.

Although barrier theorems are of theoretical nature, they have important implications in the design of numerical algorithms, elucidating the cost (in order or in numerical complexity) that is incurred by various stability conditions. Moreover, stability barriers can be breached - either by using more complicated finite difference models (e.g. multi-stage instead of one-stage) or by relaxing stability requirements (e.g. from A-stability to $A(\alpha)$-stability). Barrier theorems provide an insight into this task.

We mention in passing a different type of barrier theorems, that ensue when, instead of stability, one demands monotonicity. Although the purpose of monotonicity may vary - contractivity for ordinary differential equations [26], preservation of positivity for the heat equation [1] and avoidance of outflow from shocks for nonlinear hyperbolic conservation laws [24,32] - the barrier on order, in different situations and by different techniques, is invariably one. This is hardly satisfactory for many purposes, hence much research on relaxing monotonicity, e.g. by using TVD techniques for conservation laws [7] or by implementing absolutely monotone approximations to ordinary differential equations [22].

2. A BARRIER FOR EXPLICIT SEMI-DISCRETISATIONS

In the present section we consider numerical approximations that occur when the advection equation

$$\frac{\partial}{\partial t}u(x,t) = \frac{\partial}{\partial x}u(x,t), \quad -\infty < x < \infty, \quad t \geq 0,$$
$$u(x,o) = \psi(x), \quad -\infty < x < \infty, \tag{3}$$

is replaced by the infinite set of ordinary differential equations

$$U_m'(t) = \frac{1}{\Delta x}\sum_{k=-r}^{s} \alpha_k U_{m+k}(t), \quad m \in \mathbb{Z}, \quad t \geq 0, \tag{4}$$

$$U_m(0) = \psi(m\Delta x), \quad m \in \mathbb{Z},$$

where $U_m(t)$ approximates $u(m\Delta x,t)$, $m \in \mathbb{Z}$, $t \geq 0$. Equation (3) is important as a model for hyperbolic conservation laws.

Given a differentiable function $f(x,t)$, we define the <u>shift operator</u> by

$$E_x f(x,t) = f(x+\Delta x,t)$$

and denote the <u>differential operators</u> with respect to t and x by D_t and D_x respectively. Hence (3) gives

$$(D_t - D_x)\, u(x,t) = 0,$$

while (4) yields

$$(D_t - \frac{1}{\Delta x}\sum_{k=-r}^{s} \alpha_k E_x^k)U_m(t) = 0.$$

Let $e_m(t) := u(m\Delta x,t) - U_m(t)$, $m \in \mathbb{Z}$, be the error at $x = m\Delta x$. Hence

$$(D_t - \frac{1}{\Delta x}\sum_{k=-r}^{s} \alpha_k E_x^k)e_m(t) = (D_t - \frac{1}{\Delta x}\sum_{k=-r}^{s} \alpha_k E_x^k)u(m\Delta x,t) =$$

$$= (D_x - \frac{1}{\Delta x}\sum_{k=-r}^{s} \alpha_k E_x^k)u(m\Delta x,t), \quad m \in \mathbb{Z}, \quad t \geq 0.$$

Since, subject to analyticity,

$$D_x = \frac{1}{\Delta x}\ln E_x,$$

and $E_x = I + O(\Delta x)$, $0 < \Delta x << 1$, where I is the identity operator, it follows that (4) is of order p if and only if

$$G(z) := \sum_{k=-r}^{s} \alpha_k z^k = \ln z + c(z-1)^{p+1} + O(|z-1|^{p+2}), \quad c \neq 0, \quad z \to 1. \tag{5}$$

Stability (in the L_2 norm) follows by standard Fourier analysis: the von Neumann criterion is

$$\operatorname{Re} G(e^{i\theta}) \leq 0, \qquad -\pi \leq \theta \leq \pi . \tag{6}$$

Consequently, in approximation-theoretic setting the stability barrier problem is equivalent to maximising the order of approximation of $\ln z$ by a Laurent polynomial of degree $\{r,s\}$ subject to the real part of that polynomial being non-positive along the perimeter of the complex unit disc.

Strang [28] was the first to derive a stability barrier of this type: given $s = 1$, the maximal order of a stable scheme (4), regardless of $r \geq 0$, is $\min\{r+1, 2\}$. This was followed by Engquist and Osher [5], who provided the barrier $\min\{s,2\}$ for $r = 0$, $s \geq 1$.

The general barrier has been derived in [8] by order stars. It has been extended and embellished in [16,17,18,21] and [27]. At present at least four different proofs are available, each using a different kind of order star. In what follows we sketch a new proof (based of course, on yet another order star) which is probably the simplest available.

Let

$$\sigma(z) := G(z) - \ln z, \qquad z \in \mathbb{C}/(-\infty,0] .$$

The pair $\{A,D\}$, where

$$A := \{z \in \mathbb{C}/(-\infty,0]: \quad \operatorname{Re} \sigma(z) > 0\},$$
$$D := \{z \in \mathbb{C}/(-\infty,0]: \quad \operatorname{Re} \sigma(z) < 0\},$$

is called the <u>order star</u> (of the second kind [10]) of σ. Note that we can - and duly do - extend both A and D to $\mathbb{C}/\{0\}$ by a limiting process - although σ is discontinuous along the branch cut $(-\infty,0)$, $\operatorname{Re}\sigma$ is continuous there!

The following four propositions provide all the necessary information on the order star. Only Proposition 3 is new, hence its proof is given in full.

PROPOSITION 1. Given that the scheme (4) is of order p, the point $z=1$ is approached by $p+1$ sectors of A and $p+1$ sectors of D, each with an equal asymptotic angle of $\pi/(p+1)$.

Proof. By considering the asymptotic behaviour of σ about $z=1$ [10].

PROPOSITION 2. The scheme (4) is stable if and only if $A \cap \{z \in \mathbb{C} : |z| = 1\} = \emptyset$.

Proof. Follows at once from (6) and the definition of A.

We call the connected components of A and D A-regions and D-regions respectively. A positively-oriented portion of the boundary of an A-region which forms a closed curve on the Riemann sphere (so that ∞ can be included in the discussion) is called an A-loop if the underlying A-region lies in the interior of this curve. We use similar terminolgy for positively-oriented closed curves along a boundary of a D-region. Note that, by Proposition 1, $z=1$ lies on A-loops and D-loops.

PROPOSITION 3. (a) Between any two crossings of $z=1$, a D-loop must pass through either zero or infinity; (b) Between any two crossings of $z=1$, an A-loop must pass through either zero or infinity or the branch cut $(-\infty, 0)$.

Proof. By [17], $\operatorname{Im}\sigma(z)$ monotonically increases along D-loops and monotonically decreases along A-loops away from infinity and $(-\infty, 0]$. Moreover, crossing $(-\infty, 0)$ from top to bottom increases $\operatorname{Im}\sigma(z)$, since

$$\lim_{\varepsilon\to 0+} \{\sigma(x+i\varepsilon) - \sigma(x-i\varepsilon)\} = 2\pi i, \quad x < 0.$$

The proposition follows, since the total variation of $\operatorname{Im}\sigma(z)$ along a loop is zero.

Note that only a single A-loop that crosses $z=1$ may also cross $(-\infty,0)$ - otherwise a D-loop that crosses $z=1$ will be separated from both 0 and ∞.

PROPOSITION 4. (a) If $s \geq 1$ then s sectors of A and s sectors of D approach infinity, while if $s=0$ then $\infty \in D$; (b) If $r \geq 1$ then r sectors of A and r sectors of D approach the origin, while if $r=0$ then $0 \in A$.

Proof. By considering the asymptotic behaviour of σ as $|z| \to \infty$ or $z \to 0$, respectively.

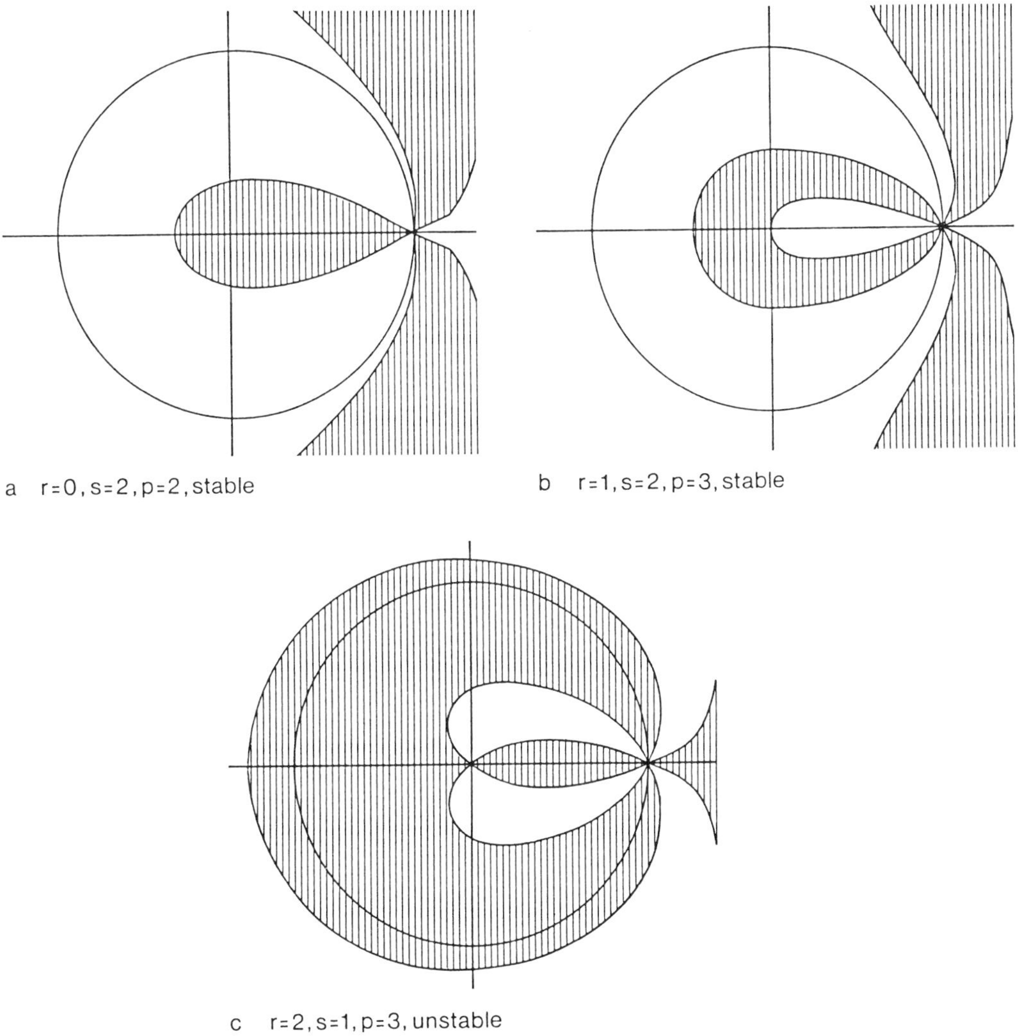

Figure 1. Order stars for schemes (4). The set A is denoted by the shaded area.

Figure 1 depicts three order stars, that illustrate our four propositions.

We are now in position to prove our barrier theorem:

THEOREM 1. The order of a stable scheme (4) may not exceed $\min\{r+s, 2r+2, 2s\}$.

Proof. The inequality $p \leq r+s$ is elementary, by counting the number of degrees of freedom. To prove $p \leq 2\min\{r+1, s\}$ we denote by M_0 and M_1 the number of sectors of A that adjoin $z=1$ from within and from without the unit disc, respectively. Note that, by stability and Proposition 2, $|z|=1$ separates between A-regions, hence M_0 and M_1 are well-defined.

It follows at once from Proposition 1 that

$$p = M_0 + M_1 - 1, \qquad M_0 - 1 \leq M_1 \leq M_0 + 1 .$$

Therefore

$$p \leq 2\min\{M_0, M_1\} . \tag{7}$$

There are clearly at most $r+1$ A-loops that cross $z=1$ and lie inside the unit disc -r that adjoin the origin and one that crosses (-1,0) - cf. Propositions 2, 3 and 4. Hence, since each sector of A at $z=1$ accounts for an A-loop, $M_0 \leq r+1$ and, by (7), $p \leq 2(r+1)$.

To obtain $p \leq 2s$ we turn our attention to the exterior of the unit disc. No A-loop there that adjoins $z=1$ may cross $(-\infty,-1)$ - otherwise, since A is symmetric with respect to the real axis, Propositions 2 and 3 will imply instability. Hence, by Proposition 4, $M_1 \leq s$ and the desired inequality follows from (7).

The barrier of Theorem 1 is attainable for every choice of $r \geq 0$ and $s \geq 0$ [8].

3. BARRIERS FOR ORDINARY DIFFERENTIAL EQUATIONS

Linear multistep-multiderivative (but not multistage) schemes for the equation (2) can be readily analysed by approximation-theoretic methods: the purpose of such schemes being to evaluate the effect of the shift operator (i.e. advancing the solution in time), the differential operator being given by the differential equation. Since the shift operator is, formally, the exponential of the differential operator, which is scaled by the step length, the theoretical framework is that of approximating the exponential function by rational or, more generally, algebraic functions.

Given

$$\rho(w) := \sum_{k=0}^{m} \alpha_k w^k ;$$

$$\sigma(w) := \sum_{k=0}^{m} \beta_k w^k ,$$

the barrier problem for the multistep scheme (1) is maximising p such that

$$\rho(e^z) - z\sigma(e^z) = O(|z|^{p+1}), \qquad z \to 0 ,$$

subject to the satisfaction of the root condition by ρ : all zeros of ρ need be in the closed complex unit disc and all zeros on the unit circle need be simple.

Dahlquist [3] proved that $p \leq 2[\frac{1}{2}(m+1)]$ by elementary methods. Alongside several more recent proofs of that barrier there is one based on order stars [14]. The barrier has been generalised to cater for higher derivatives by Reimer [23] and, more recently and comprehensively, by Jeltsch and Nevanlinna [19].

An allied problem is that of backward difference formulae (BDF). σ is restricted to the form $\sigma(w) = \beta_m w^m$, the order is $p = m+1$ and the "inverse" barrier consists of finding the range of $m \geq 1$ such that ρ satisfies the root condition. This was solved by Cryer [2], who showed that $1 \leq m \leq 6$. A shorter proof has recently been published by Hairer and Wanner [6], who extensively use approximation-theoretic techniques.

Different types of barriers occur when A-stability, rather than zero-stability (which is equivalent to the root condition for consistent schemes) is required. The results are more general, since they pertain to multistage schemes as well. Again, the exponential function is approximated by a solution of an algebraic function, but now the approximation need be bounded in modulus by one in the complex left half-plane.

Again, it was Dahlquist who provided the first result of this type [4]: the order of an A-stable multistep method (1) is at most 2 and the second-order A-stable method with the least error constant is the (one-step) trapezoidal rule. The generalisation of this barrier to the multiderivative (and multistage) framework was a long-standing conjecture of Daniel and Moore. Its solution [31] was among the first triumphs of order stars: the order of an A-stable multistep ℓ-derivative method is at most 2ℓ and the 2ℓ-th-order method with the least error constant is the (one-step) $[\ell/\ell]$ Padé-Obrechkoff scheme.

The attainment of the barrier by one-step methods motivates the whole subject of rational approximation to the exponential. Let $\pi_{m/n}[z]$ denote the set of all rational functions of the form $P(z)/Q(z)$, where $\deg P = m$, $\deg Q = n$ and $Q(0) = 1$. For each $m,n \geq 0$ there exists a unique $R_{m/n} \in \pi_{m/n}[z]$ such that

$$R_{m/n}(z) = e^z + O(|z|^{m+n+1}), \qquad z \to 0.$$

It is called the [m/n] <u>Padé approximation</u> to the exponential. A function $R \in \pi_{m/n}[z]$ is said to be <u>A-acceptable</u> if $|R(z)| \leq 1$ for every $z \in \mathbb{C}$ such that $\mathrm{Re}\, z \leq 0$ - the connection between A-stable methods and A-acceptable rational functions needs no further elaboration.

It has been known for some time that the choice $0 \leq m \leq n \leq m+2$ always yields an A-acceptable Padé approximation and Ehle conjectured that no other Padé approximation may be A-acceptable. This has been proved by Wanner, Hairer and Nørsett [31] and was the original reason for the introduction of order stars.

The order of the approximation can be relaxed from m+n and replaced by interpolation to the exponential along the negative semi-axis - this is important in certain applications. The second Ehle conjecture stated that all such A-acceptable rational functions are obtained when $0 \leq m \leq n \leq m+2$ and the order at the origin is at least $2n-2$. It has been proved (with order stars) by Iserles and Powell [15].

Many further results, beyond the scope of this review, are available: interpolation points may be complex conjugate, the number of distinct poles may be restricted etc. In all these order stars are instrumental.

4. BARRIERS FOR PARTIAL DIFFERENTIAL EQUATIONS

Theorem 1 has already provided an example of a barrier for the advection equation (3). It can be generalised to cater for implicit semi-discretised schemes

$$\sum_{k=-R}^{S} \beta_k U'_{m+k}(t) = \frac{1}{\Delta x} \sum_{k=-r}^{s} \alpha_k U_{m+k}(t), \quad m \in \mathbb{Z}, \quad t \geq 0 .$$

Let

$$G(z) := \frac{\sum_{k=-r}^{s} \alpha_k z^k}{\sum_{k=-R}^{S} \beta_k z^k} .$$

Again, order p is equivalent to

$$G(z) = \ln z + c(z-1)^{p+1} + O(|z-1|^{p+2}), \quad c \neq 0, \quad z \to 1,$$

whereas stability rests on two criteria: the von Neumann condition (6), as well as the pole condition - out of the S+R poles of G in $\mathbb{C}/\{0\}$ R must reside within and S without the complex unit disc [16]. The stability barrier has been derived by Iserles and Williamson [17],

$$p \leq \min\{ r+s+R+S, \ 2(r+R+2), \ 2(s+S)\} . \tag{8}$$

It is attainable for a wide range of choices of r,s,R and S [12,17].

Further insight has been provided by Jeltsch and Strack [21,27], who obtained least bounds on the size of the error constant, subject to stability and to equality in (8). It transpires that, even in the explicit case, such bounds are not always attainable, since they correspond to an unstable method which can be "approached" arbitrarily close by stable schemes with larger error constant.

Order stars are also essential in deriving stability barriers for fully discretised schemes for the advection equation, of the form

$$\sum_{k=-R}^{S} \delta_k(\mu) U_{m+k}^{n+1} = \sum_{k=-r}^{s} \gamma_k(\mu) U_{m+k}^{n}, \quad m \in \mathbb{Z}, \quad n \geq 0, \tag{9}$$

where $\mu = \Delta t/\Delta x$ is the Courant number. Instead of approximating $\ln z$, we consider approximations to z^{μ} about $z=1$. Iserles and Strang [16] proved that the bound (8) remains valid in the present framework when $\mu \to 0+$ and subject to the condition that for $\mu = 0$ the difference scheme reduces to identity. Smit [25] showed that, for explicit schemes, the bound holds for every $\mu \in (0,1)$ - again, subject to the same condition. Finally, recently both Jeltsch and Smit [20] and Iserles [13] demonstrated that (8) remains valid for every scheme (9) and every μ in (0,1), with no restrictions.

Mindful of the theory of Section 3, we now extend the discussion to multistep (or, in partial differential equations' parlance, multilevel) schemes of the form

$$\sum_{\ell=0}^{L} \sum_{k=-r}^{s} \eta_{\ell,k}(\mu) U_{m+k}^{n+\ell} = 0, \quad m \in \mathbb{Z}, \quad n \geq 0.$$

Now z^{μ} is approximated by a branch of an algebraic function. Strang and Iserles [29] showed, by using standard function-theoretic techniques, that, subject to stability, the order may not exceed 2(r+s), independent of the number of steps. More powerful barriers, that concentrate on multistep methods of a more specific form, were presented in [9] and will be the theme of forthcoming papers by R. Jeltsch and by the present author. Order stars are central to this work. For example, in [9] all maximal-order, stable, conservative, explicit two-step methods are characterised by considering approximations to $\sinh \mu z$ by a linear combination of $\sinh z$, $\sinh 2z, \ldots,$ $\sinh \ell z$ and applying the appropriate order star.

An aspect of numerical schemes for the advection equation which provides important information on their performance is group velocity - cf. the review of Trefethen [30]. It can be incorporated into the order-stability framework. Again, this is best done in an approximation-theoretical setting. Alas, although the behaviour of group velocity can be interpreted by the geometry of the underlying order star, this has not proved, so far, to be very useful [12].

The discussion of influence of stability on order need not be confined to the advection equation. A discussion of the heat equation

$$\frac{\partial}{\partial t} u(x,t) = \frac{\partial^2}{\partial t^2} u(x,t), \quad x \in \mathbb{R}, \quad t \geq 0,$$

with an appropriate initial condition, leads to a surprising "anti-barrier": the maximal order of a stable scheme may exceed the number of degrees of freedom [11]! Specifically, let us consider the scheme

$$\sum_{k=-r}^{r} \delta_k(\mu) U_{m+k}^{n+1} = \sum_{k=-r}^{r} \gamma_k(\mu) U_{m+k}^{n}, \quad m \in \mathbb{Z}, \quad n \geq 0, \tag{10}$$

where $\mu = \Delta t/(\Delta x)^2$ is constant - hence the error (and, by implication, the order) can be expressed in terms of Δx only. A typical method is the Crank-Nicolson scheme which, with $r = 1$, is stable and of order 3. Although Crank-Nicolson is very popular in applications, in fact one can do better: there exists a fifth-order stable method with $r = 1$. The general pattern emerges upon translation to approximation theory. (10) now corresponds to a Laurent polynomial approximation to $\exp(\mu(\ell n\, z)^2)$ about $z = 1$ and order stars prove that, for every $r \geq 1$, there exists a unique method of the maximal order $4r + 1$ and that this method is stable [11].

REFERENCES

1. BOLLEY, C. & CROUZEIX, M. Conservation de la positivité lors de la discrétisation des problèmes d'évolution paraboliques, <u>RAIRO Analyse Numérique 12</u> (1978) 237-245.

2. CRYER, C.W. On the instability of high order backward-difference multistep methods, <u>BIT 12</u> (1972) 17-25.

3. DAHLQUIST, G. Convergence and stability in the numerical integration of ordinary differential equations, <u>Math. Scand. 4</u> (1956) 33-53.

4. DAHLQUIST, G. A special stability problem for linear multistep methods, <u>BIT 3</u> (1963) 27-43.

5. ENGQUIST, B. & OSHER, S. One-sided difference approximations for non-linear conservation laws, <u>Math. Comp. 36</u> (1981) 321-352.

6. HAIRER, E. & WANNER, G. On the instability of the BDF formulas, <u>SIAM J. Num. Analysis 20</u> (1983) 1206-1209.

7. HARTEN, A. High resolution schemes for hyperbolic conservation laws, <u>J. Comput. Phys. 49</u> (1983) 357-393.

8. ISERLES, A. Order stars and a saturation theorem for first-order hyperbolics, <u>IMA J. Num. Analysis 2</u> (1982) 49-61.

9. ISERLES, A. Generalized leapfrog methods, Univ. of Cambridge Tech. Rep. DAMTP NA8 (1984).

10. ISERLES, A. Order stars, approximations and finite differences I. The general theory of order stars, <u>SIAM J. Math. Analysis 16</u> (1985) to appear.

11. ISERLES, A. Order stars, approximations and finite differences III. Finite differences for $u_t = \omega u_{xx}$, SIAM J. Math. Analysis 16 (1985) to appear.

12. ISERLES, A. Order, stability and group velocity I. Semi-discretised schemes, Univ. of Cambridge Tech. Rep. DAMTP NA5 (1985).

13. ISERLES, A. Order and stability of fully-discretised hyperbolic finite differences, Univ. of Cambridge Tech. Rep. DAMTP NA6 (1985).

14. ISERLES, A. & NØRSETT, S.P. A proof of the first Dahlquist barrier by order stars, BIT 24 (1984) 529-537.

15. ISERLES, A. & POWELL, M.J.D. On the A-acceptability of rational approximations that interpolate the exponential, IMA J. Num. Analysis 1 (1981) 241-251.

16. ISERLES, A. & STRANG, G. The optimal accuracy of difference schemes, Trans. Am. Math. Soc. 277 (1983) 779-803.

17. ISERLES, A. & WILLIAMSON, R.A. Stability and accuracy of semi-discretized finite difference methods, IMA J. Num. Analysis 4 (1984) 289-307.

18. JELTSCH, R. Stability and accuracy of difference schemes for hyperbolic problems, RWTH Aachen Tech. Rep. 27 (1984).

19. JELTSCH, R. & NEVANLINNA, O. Dahlquist's first barrier for multistage multistep formulas, BIT 24 (1984) 538-555.

20. JELTSCH, R. & SMIT, J.H. Bounds for the accuracy of difference methods for hyperbolic differential equations, RWTH Aachen Tech. Rep. 31(1985).

21. JELTSCH, R. & STRACK, K.-G. Accuracy bounds for semidiscretizations of hyperbolic problems, Math. Comp. 44 (1985) to appear.

22. KRAAIJEVANGER, J.F.B.M. Absolute monotonicity of polynomials occurring in the numerical solution of initial value problems, Univ. of Leiden Tech. Rep. 84-01 (1984).

23. REIMER, M. Finite difference forms containing derivatives of higher order, SIAM J. Num. Analysis 5 (1968) 725-738.

24. ROE, P.L. Numerical algorithms for the linear wave equation, Royal Aircraft Est. Bedford Tech. Rep. (1981).

25. SMIT, J.H. Order stars and the optimal accuracy of stable, explicit difference schemes, Univ. of Stellenbosch Tech. Rep. (1984).

26. SPIJKER, M.N. On the relation between stability and contractivity, Univ. of Leiden Tech. Rep. 84-03 (1984).

27. STRACK, K.-G. Stability and accuracy for implicit semidiscretizations of hyperbolic problems, RWTH Aachen Tech. Rep. 28 (1984).

28. STRANG, G. Accurate partial difference methods II. Non-linear problems, Numer. Math. 6 (1964) 37-46.

29. STRANG, G. & ISERLES, A. Barriers to stability, SIAM J. Num. Analysis 20 (1983) 1251-1257.

30. TREFETHEN, L.N. Group velocity in finite difference schemes, SIAM Rev. 24 (1982) 113-136.

31. WANNER, G., HAIRER, E. & NØRSETT, S.P. Order stars and stability theorems, BIT 18 (1978) 475-489.

32. WIDLUND, O.B. On Lax's theorem on Friedrichs-type finite difference schemes, Comm. Pure Appl. Math. 24 (1971) 117-123.

A. Iserles
King's College
University of Cambridge
Cambridge CB2 1ST
England

C VAN LOAN

Parallel algorithms for constrained and unconstrained least squares problems

1. INTRODUCTION

During the past few years we have been concerned with the parallel computation of the singular value decomposition (SVD). In the SVD we are given an m-by-n matrix A and compute orthogonal U (m-by-m) and V (n-by-n) such that

$$U^T A V = \Sigma = \mathrm{diag}(\sigma_1, \ldots, \sigma_n) \quad . \qquad (1.1)$$

The σ_i are the singular values and U and V can be chosen so that they satisfy' $\sigma_1 \geqq \cdots \geqq \sigma_n \geqq 0$. The SVD is one of the most important matrix decompositions in numerical linear algebra and many of its numerous applications are detailed in Golub and Van Loan [3].

Perhaps the leading application of the SVD is the least square problem

$$\min \|Ax - b\|_2 \qquad A \in R^{m\times n}, \quad b \in R^m \qquad (1.2)$$

especially when A is rank degenerate. Being able to solve this problem in real-time is of interest in several signal processing applications. Unfortunately, conventional computers are unable to handle this task because m and n are too big for the alotted time that one has to get the solution x. Thus, it is not surprising that the real-time signal processing community has become quite interested in special purpose architectures that can solve least square problems in parallel.

In §2 we describe one such architecture-a systolic array-that has been proposed for computing the SVD. The underlying algorithm is based on a Jacobi procedure. We show how a few simple extensions of the procedure can widen the class of problems that the architecture can handle. In particular, we present what we call a Jacobi submatrix SVD procedure that can be used to solve constrained least square problems of the variety

$$\min_{Bx = d} \|Ax - b\|_2 \qquad A \in R^{m\times n}, \; b \in R^m, \; B \in R^{p\times n}, \; d \in R^p \qquad (1.3)$$

In §3 we discuss a block analogue of the Jacobi SVD procedure that is suitable for the case of a few loosely coupled but powerful processors.

One problem with the SVD is that it is hard to update. That is, if we have the SVD of a matrix A and then want the SVD of a rank one perturbation of A, we pretty much have to start from scratch. This makes the QR factorization much more attractive in least squares settings where updating is required. For example, in a beamforming application that we are aware of, we must solve the problem (1.3) for a fixed A but with many different B. In §4 we show how to extend some parallel QR factorization procedures so that they can handle this case.

Many of the techniques that we discuss in this paper are detailed elsewhere. Our intention is to give an overview of work in this area and to impress upon the reader the need for parallel algorithms that are simple enough to be implemented on a special purpose device but general enough so that they can solve more than one narrowly defined problem.

2. SUBMATRIX SVD USING JACOBI ROTATIONS

Assume for the moment that A is a square n-by-n real matrix. In the Jacobi approach to computing the SVD, A is made increasingly diagonal through updates of the form

$$A := J(p,q,\theta_1)^T A \, J(p,q,\theta_2) \; . \tag{2.1}$$

Here, $J(p,q,\theta)$ is a Jacobi rotation in the (p,q) plane, i.e., $J(p,q,\theta) = (z_{ij})$ is the identity everywhere except

$$\begin{bmatrix} z_{pp} & z_{pq} \\ z_{qp} & z_{qq} \end{bmatrix} = \begin{bmatrix} \cos(\theta) & \sin(\theta) \\ -\sin(\theta) & \cos(\theta) \end{bmatrix} .$$

For a given index pair (p,q) in (2.1) it is possible to choose θ_1 and θ_2 such that

$$\begin{bmatrix} \cos(\theta_1) & \sin(\theta_1) \\ -\sin(\theta_1) & \cos(\theta_1) \end{bmatrix}^T \begin{bmatrix} a_{pp} & a_{pq} \\ a_{qp} & a_{qq} \end{bmatrix} \begin{bmatrix} \cos(\theta_2) & \sin(\theta_2) \\ -\sin(\theta_2) & \cos(\theta_2) \end{bmatrix} = \begin{bmatrix} d_1 & 0 \\ 0 & d_2 \end{bmatrix} \tag{2.2}$$

This amounts to solving a 2-by-2 SVD problem. By choosing the rotations in in this fashion and updating A as in (2.1), the sum of the squares of A's off-diagonal elements is reduced by the amount $a_{pq}^2 + a_{qp}^2$. If the rotation indices (p,q) are chosen properly then the matrix A converges to the diagonal matrix of singular values. Formally, if

$$\text{off}(A) = \text{sqrt}\left(\sum_{i \neq j} a_{ij}^2 \right)$$

ε = given tolerance, e.g., unit roundoff

$\tau = \varepsilon \| A \|_F / n$

$\{(p_i, q_i)\}_{i=1}^{N}$ be an ordering of $\{(p,q) \mid 1 \leq p < q \leq n\}$, $N = \frac{n(n-1)}{2}$

then the following algorithm terminates:

```
Do While ( off(A) > ε ||A||_F )
|  For i = 1:N
|  |  (p,q) := (p_i,q_i)
|  |  If  a_pq^2 + a_qp^2 > τ^2                    (2.3)
|  |     then
|  |        Solve the 2-by-2 SVD problem (2.2).
|  -        A := J(p,q,θ_1)^T A J(p,q,θ_2)
```

A proof of termination may be found in [6]. The parameter τ is called the threshold parameter. Its presence in the algorithm ensures that we only perform "worthwhile" rotations. Thealgorithm may not terminate if τ is set to zero. How to solve the 2-by-2 subproblems is discussed in [2].

Perhaps the most interesting aspect of (2.3) is the ordering. The cyclic-by-row ordering is most often discussed in elementary texts:

$$(1,2),(1,3),\ldots,(1,n),(2,3),\ldots(2,n),\ldots.(n-1,n) .$$

For most reasonable orderings, (2.3) will converge after a small number of sweeps. (A single pass through the For-loop is called a sweep.)

The most exciting development in the last few years concerning Jacobi methods was the discovery of the "parallel" ordering in [1]. For n = 8, the parallel ordering is as follows:

I.	(1,2)	(3,4)	(5,6)	(7,8)
II.	(1,4)	(2,6)	(3,8)	(5,7)
III.	(1,6)	(4,8)	(2,7)	(3,5)
IV.	(1,8)	(6,7)	(4,5)	(2,3)
V.	(1,7)	(5,8)	(3,6)	(2,4)
VI.	(1,5)	(3,7)	(2,8)	(4,6)
VII.	(1,3)	(2,5)	(4,7)	(6,8)

The table should be read left-to-right, top-to-bottom. Notice that the rotations prescribed by a given row are "nonconflicting". That is, they involve independent subproblems. A parallel implementation of (2.3) might read:

Solve all the subproblems associated with rotation set I.
Perform the row updates associated with each subproblem.
Perform the column updates associated with each subproblem.
Solve all the subproblems associated with rotation set II.

⋮

The fact that the subproblems and updates are independent, means that they can be performed concurrently.

In [2] we show how an array of processors of the form

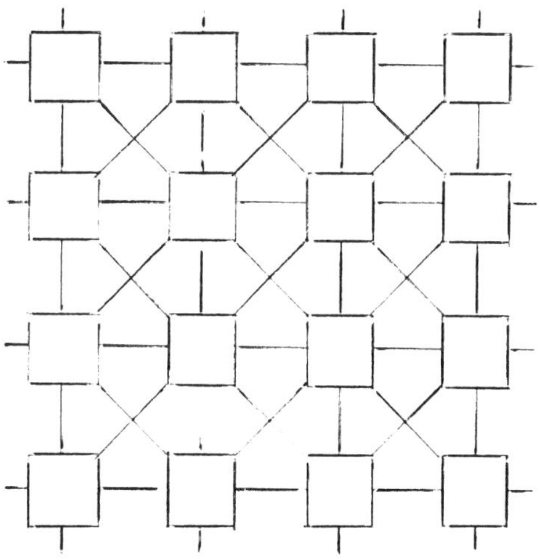

(2.4)

can be used to carry out a sweep in the parallel ordering in O(n) time steps. The diagonal processors solve 2-by-2 SVD problems while the off-diagonal processors perform the 2-by-2 matrix multiplications necessary to carry out the updates $A := J(p,q,\theta_1)^T A\, J(p,q,\theta_2)$.

At this point we'd like to dwell on an interesting practical question. Suppose we had an array of processors capable of solving 8-by-8 SVD problems. How could it be used to solve, say, a 6-by-5 problem? An obvious course of action might be to pad A with zeroes to make it 8-by-8:

$$\bar{A} = \begin{matrix} x & x & x & x & x & o & o & o \\ x & x & x & x & x & o & o & o \\ x & x & x & x & x & o & o & o \\ x & x & x & x & x & o & o & o \\ x & x & x & x & x & o & o & o \\ x & x & x & x & x & o & o & o \\ o & o & o & o & o & o & o & o \\ o & o & o & o & o & o & o & o \end{matrix} = \begin{bmatrix} A & 0 \\ 0 & 0 \end{bmatrix} \tag{2.5}$$

One would then expect the array to compute the SVD

$$\begin{bmatrix} U & 0 \\ 0 & I_2 \end{bmatrix}^T \begin{bmatrix} A & 0 \\ 0 & 0 \end{bmatrix} \begin{bmatrix} V & 0 \\ 0 & I_3 \end{bmatrix} = \begin{bmatrix} \Sigma & 0 \\ 0 & 0 \end{bmatrix}$$

whereupon $U^T AV = \Sigma$. Trouble can arise, however, if A has a nontrivial nullspace for then the orthogonal transformations in (2.5) need not be in direct sum form. In this case we say that the "genuine" nullspace of A intermingles with the "artificial" nullspace of $\bar{A}$ that arises because of the bordering of zeroes. The nullspace of A can be retrieved from the computed nullspace of $\bar{A}$, but this is inconvenient. More importantly, the accuracy of A's computed singular vectors that are associated with small singular values is degraded by the intermingling effect.

Fortunately, these difficulties can be avoided if the 2-by-2 subproblems are carefully solved. The idea is to avoid the intermingling of A and its bordering of zeroes. For example, if we are doing a (4,6) rotation, then the associated 2-by-2 SVD problem can be solved in two ways:

$$\begin{bmatrix} \cos(\theta_1) & \sin(\theta_1) \\ -\sin(\theta_1) & \cos(\theta_1) \end{bmatrix}^T \begin{bmatrix} x & 0 \\ x & 0 \end{bmatrix} \begin{bmatrix} 1 & 0 \\ 0 & 1 \end{bmatrix} = \begin{bmatrix} x & 0 \\ 0 & 0 \end{bmatrix}$$

$$\begin{bmatrix} \cos(\theta_1) & \sin(\theta_1) \\ -\sin(\theta_1) & \cos(\theta_1) \end{bmatrix}^T \begin{bmatrix} x & 0 \\ x & 0 \end{bmatrix} \begin{bmatrix} 0 & 1 \\ -1 & 0 \end{bmatrix} = \begin{bmatrix} 0 & 0 \\ 0 & x \end{bmatrix}$$

Notice that if the second choice of sines and cosines is taken, then the zero-nonzero structure of the original $\ddot{A}$ is destroyed insofar as columns 4 and 6 are swapped. Clearly, we should set $\theta_2 = 0$ when $(p,q) = (4,6)$. More generally, if we set $\theta_1 = 0$ whenever $q \geqq 7$ and $\theta_2 = 0$ whenever $q \geqq 6$ then one can prove that the "direct sum" SVD (2.6) ensues.

By being careful at the 2-by-2 subproblem level, as when avoiding the intermingling problem, one can develop Jacobi procedures that compute the SVD of a designated submatrix. This is an important capability as it widens the class of problems that the SVD array can solve. As a case in point, consider the constrained least squares problem (1.3). Suppose that we compute orthogonal U_1, U_2, and V such that

$$\begin{bmatrix} U_1 & 0 \\ 0 & U_2 \end{bmatrix}^T \begin{bmatrix} B \\ A \end{bmatrix} V = \begin{bmatrix} D_B & 0 \\ A_1 & D_A \end{bmatrix} \tag{2.6}$$

where D_B (p-by-p) and D_A (m-by-(n-p)) are diagonal. This can be accomplished by computing the SVD

$$B = U_1 \, [\, D_B \quad 0 \,] \, V_1^T$$

and then computing the SVD of the last n-p columns of AV_1 . With these reductions (1.3) transforms to the easily solved problem

$$\begin{aligned} &\min \; \| D_A y_2 - (U_2^T b - A_1 y_1) \|_2 \\ &D_B y_1 = U_1^T d \end{aligned}$$

whereupon

$$x = V \begin{bmatrix} y_1 \\ y_2 \end{bmatrix} \qquad y_1 \in R^p \ , \ y_2 \in R^{n-p} \quad .$$

Let's see how a slight modification of algorithm (2.3) can be used to compute the SVD-like decomposition (2.6).

Suppose we have an 8-by-8 array and that $A \in R^{6x5}$ and $B \in R^{2x5}$. We initialize the processor array as follows:

$$\begin{matrix} b & b & b & b & b & o & o & o \\ b & b & b & b & b & o & o & o \\ a & a & a & a & a & o & o & o \\ a & a & a & a & a & o & o & o \\ a & a & a & a & a & o & o & o \\ a & a & a & a & a & o & o & o \\ a & a & a & a & a & o & o & o \\ o & o & o & o & o & o & o & o \end{matrix}$$

Here, a (b) denotes an arbitrary nonzero element of A (B). The first calculation to be performed involves the SVD of the submatrix defined by rows 1 and 2 and columns 1 through 5. Noting that submatrices can be designated by bit array we can specify the above mentioned submatrix as follows:

$$\begin{matrix} R & R & R & R & R & N & N & N \\ R & R & R & R & R & N & N & N \\ N & N & N & N & N & N & N & N \\ N & N & N & N & N & N & N & N \\ N & N & N & N & N & N & N & N \\ N & N & N & N & N & N & N & N \\ N & N & N & N & N & N & N & N \\ N & N & N & N & N & N & N & N \end{matrix}$$

Here, "R" indicates a submatrix entry while "N" indicates an entry "outside" the submatrix. When the Jacobi SVD procedure is applied to the 8-by-8 matrix there is a 2-by-2 bit array associated with each 2-by-2 SVD subproblem. There are four possibilities:

$$\underset{(a)}{\begin{bmatrix} R & R \\ R & R \end{bmatrix}} \qquad \underset{(b)}{\begin{bmatrix} R & R \\ N & N \end{bmatrix}} \qquad \underset{(c)}{\begin{bmatrix} R & N \\ N & N \end{bmatrix}} \qquad \underset{(d)}{\begin{bmatrix} N & N \\ N & N \end{bmatrix}}$$

If the 2-by-2 subproblems in each of these cases is solved as follows,

$$(a) : \quad \begin{bmatrix} c_1 & s_1 \\ -s_1 & c_1 \end{bmatrix}^T \begin{bmatrix} x & x \\ x & x \end{bmatrix} \begin{bmatrix} c_2 & s_2 \\ -s_2 & c_2 \end{bmatrix} = \begin{bmatrix} x & 0 \\ 0 & x \end{bmatrix}$$

$$(b):\ \begin{bmatrix}1 & 0\\ 0 & 1\end{bmatrix}^T \begin{bmatrix}x & x\\ x & x\end{bmatrix} \begin{bmatrix}c_2 & s_2\\ -s_2 & c_2\end{bmatrix} = \begin{bmatrix}x & 0\\ x & x\end{bmatrix}$$

$$(c):\ \begin{bmatrix}1 & 0\\ 0 & 1\end{bmatrix}^T \begin{bmatrix}x & x\\ x & x\end{bmatrix} \begin{bmatrix}1 & 0\\ 0 & 1\end{bmatrix} = \begin{bmatrix}x & x\\ x & x\end{bmatrix}$$

$$(d)\ \begin{bmatrix}1 & 0\\ 0 & 1\end{bmatrix}^T \begin{bmatrix}x & x\\ x & x\end{bmatrix} \begin{bmatrix}1 & 0\\ 0 & 1\end{bmatrix} = \begin{bmatrix}x & x\\ x & x\end{bmatrix}$$

then after a sufficient number of sweeps the array is transformed to

x	o	o	o	o	o	o	o
o	x	o	o	o	o	o	o
x	x	x	x	x	o	o	o
x	x	x	x	x	o	o	o
x	x	x	x	x	o	o	o
x	x	x	x	x	o	o	o
x	x	x	x	x	o	o	o
o	o	o	o	o	o	o	o

Notice that the 2-by-2 subproblems are solved in a way that avoids the mingling of "N" entries and "R" entries.

The second step in the computation of (2.6) is to compute the SVD of the submatrix defined by columns 3, 4, and 5 and rows 3 through 7 . For this phase of the calculation the following bit array is pertinant:

N	N	N	N	N	N	N	N
N	N	N	N	N	N	N	N
N	N	R	R	R	N	N	N
N	N	R	R	R	N	N	N
N	N	R	R	R	N	N	N
N	N	R	R	R	N	N	N
N	N	R	R	R	N	N	N
N	N	N	N	N	N	N	N

There are five "types" of 2-by-2 subproblems in this case. If they are solved as follows:

$$\begin{bmatrix} R & R \\ R & R \end{bmatrix} : \quad \begin{bmatrix} c_1 & s_1 \\ -s_1 & c_1 \end{bmatrix}^T \begin{bmatrix} x & x \\ x & x \end{bmatrix} \begin{bmatrix} c_2 & s_2 \\ -s_2 & c_2 \end{bmatrix} = \begin{bmatrix} x & 0 \\ 0 & x \end{bmatrix}$$

$$\begin{bmatrix} R & N \\ R & N \end{bmatrix} : \quad \begin{bmatrix} c_1 & s_1 \\ -s_1 & c_1 \end{bmatrix}^T \begin{bmatrix} x & x \\ x & x \end{bmatrix} \begin{bmatrix} 1 & 0 \\ 0 & 1 \end{bmatrix} = \begin{bmatrix} x & x \\ 0 & x \end{bmatrix}$$

$$\begin{bmatrix} R & N \\ N & N \end{bmatrix} : \quad \begin{bmatrix} 1 & 0 \\ 0 & 1 \end{bmatrix}^T \begin{bmatrix} x & x \\ x & x \end{bmatrix} \begin{bmatrix} 1 & 0 \\ 0 & 1 \end{bmatrix} = \begin{bmatrix} x & x \\ x & x \end{bmatrix}$$

$$\begin{bmatrix} N & N \\ N & R \end{bmatrix} : \quad \begin{bmatrix} 1 & 0 \\ 0 & 1 \end{bmatrix}^T \begin{bmatrix} x & x \\ x & x \end{bmatrix} \begin{bmatrix} 1 & 0 \\ 0 & 1 \end{bmatrix} = \begin{bmatrix} x & x \\ x & x \end{bmatrix}$$

$$\begin{bmatrix} N & N \\ N & N \end{bmatrix} : \quad \begin{bmatrix} 1 & 0 \\ 0 & 1 \end{bmatrix}^T \begin{bmatrix} x & x \\ x & x \end{bmatrix} \begin{bmatrix} 1 & 0 \\ 0 & 1 \end{bmatrix} = \begin{bmatrix} x & x \\ x & x \end{bmatrix}$$

then the matrix in the array converges to

$$\begin{matrix} x & o & o & o & o & o & o & o \\ o & x & o & o & o & o & o & o \\ x & x & x & o & o & o & o & o \\ x & x & o & x & o & o & o & o \\ x & x & o & o & x & o & o & o \\ x & x & o & o & o & o & o & o \\ x & x & o & o & o & o & o & o \\ o & o & o & o & o & o & o & o \end{matrix}$$

Thus, the decomposition (2.6) is achieved. The matrices U_1 and U_2 are the accumulation of all the left rotations while V is the accumulation of all the right rotations.

It is natural to wonder why algorithm (2.3) should converge with all these stipulations on how the 2-by-2 subproblems are solved. The reason is simple. If we associate a zero with each "N" entry then our "submatrix SVD" procedure is equivalent to (2.3) with no intermingling which we know converges.

The bit array idea is attractive in that it is a simple way to "reprogram" the processor array. So long as the diagonal processors in (2.4) are set up to decipher 2-by-2 bit arrays, we're able to solve more than just simple SVD problems. This is what we mean by a simple extension of a basic algorithm that widens the class of problems that it can solve.

3. BLOCK JACOBI SVD

In the parallel matrix computation area, block algorithms can sometimes be used to achieve a more favorable ratio of computation to communication. Consider the processor array (2.4). Basically, information is passed in between neighboring processors after 2-by-2 matrix computations are performed. If the processors could solve larger problems then more time might be spent computing and less (relatively) on communication.

In this connection it is important to recognize that algorithm (2.3) has a block analogue. Suppose

$$A = \begin{bmatrix} A_{11} & A_{12} & \cdots & A_{1n} \\ A_{21} & A_{22} & \cdots & A_{2n} \\ \vdots & \vdots & & \vdots \\ A_{n1} & A_{n2} & \cdots & A_{nn} \end{bmatrix}$$

where each A_{ij} is a r-by-r matrix. (We have assumed that everything is square for expository purposes.) Here's the block version of (2.3) with accumulation of U and V:

$V := I$

$U := I$

Do While ($OFF(A) > \varepsilon \| A \|_F$)

For $i = 1:N$

$(p,q) = (p_i,q_i)$

If $\| A_{pq} \|_F^2 + \| A_{qp} \|_F^2 > \tau^2$

then

Solve the 2r-by-2r SVD problem

$$\bar{U}^T \begin{bmatrix} A_{pp} & A_{pq} \\ A_{qp} & A_{qq} \end{bmatrix} \bar{V} = \bar{\Sigma}$$

$A := J(p,q,\bar{U})^T A \; J(p,q,\bar{V})$

$U := U \; J(p,q,\bar{U})$

$V := V \; J(p,q,\bar{V})$

Here, OFF($\cdot$) is defined by

$$\mathrm{OFF}(A) = \mathrm{sqrt}\left(\sum_{i \neq j} \| A_{ij} \|_F^2 \right)$$

and $J(p,q,V)$ denotes the nr-by-nr orthogonal matrix (Z_{ij}) that is the identity everywhere except that $Z_{pp} = \bar{V}_{11}$, $Z_{pq} = \bar{V}_{12}$, $Z_{qp} = \bar{V}_{21}$, and $Z_{qq} = \bar{V}_{22}$.

All of the properties of the scalar Jacobi SVD carry over to the block algorithm. However, there is some additional flexibility. For example, it turns out to be unnecessary to compute the full SVD in each subproblem. See [6] for details. We are implementing the block Jacobi SVD procedure in an environment that has several FPS 164's connected to an IBM host. We will report on that experience elsewhere.

4. A PARALLEL IMPLEMENTATION OF THE METHOD OF WEIGHTING

The last topic that we would like to cover concerns the solution of the constrained problem (1.3). It is well-known how to use the QR factorization to solve this problem:

1. $B^T = [Q_{1B} , Q_{2B}] \begin{bmatrix} R_B \\ 0 \end{bmatrix}$. (QR factorization)

2. Solve the lower triangular system $R_B^T y_1 = d$ for $y_1 \in R^p$.

3. Set $A_1 = AQ_{2B}$, compute its QR factorization and solve the unconstrained least square problem

$$\min \| A_1 y_2 - (b - AQ_{1B}y_1) \|_2$$

for $y_2 \in R^{n-p}$.

4. $x = Q_{1B}y_1 + Q_{2B}y_2$.

Note that two QR factorizations are required. These could be solved very fast using any of several QR processor arrays that have been recently proposed. See [4] for example.

The trouble with this approach is twofold. First, there is substantial communication overhead because two "calls" to the QR processor array are required. Second, in certain signal processing applications one is interested in the solution to (1.3) for many different B. In this context, one must repeat the entire computation for each B.

A way round these difficulties is to use the method of weights. In this approach the unconstrained problem

$$\min \left\| \begin{bmatrix} \lambda B \\ A \end{bmatrix} x - \begin{bmatrix} \lambda d \\ b \end{bmatrix} \right\|_2$$

is solved for a suitably large weight λ . Mathematically, the larger the weight the better x will approximate the true solution. Note that if the QR factorization of A is known then it can be updated to obtain a QR factorization of

$$C = \begin{bmatrix} \lambda B \\ A \end{bmatrix} .$$

This can be implemented very elegantly on a very slightly modified Heller-Ipsen QR array [4].

Followers of the method of weighting know that it can be unstable for large λ . However, we recently showed how to implement the technique with small, "safe" weights. A single QR factorization of the matrix C is required. Thereafter, a few iterative improvement steps are performed to obtain an accurate solution x. See [5] for details. Our work on this problem is yet another example of how research in the parallel matrix computation area can lead to better sequential algorithms.

REFERENCES

1. BRENT, R. and F. LUK. The Solution of singular value and symmetric eigenvalue problems on multiprocessor arrays, SIAM J. Sci. Stat. Comp 6 (1985) 69-84.
2. BRENT, R., F. LUK, and C. VAN LOAN Computation of the singular value decomposition using mesh-connected processors, J. VLSI and Computer Systems 1 (1985) 242-270.
3. GOLUB, G.H. and C. VAN LOAN Matrix Computations (1983) , Johns Hopkins University Press, Baltimore, Maryland.
4. HELLER, D. and I. IPSEN Systolic networks for orthogonal decompositions, SIAM J. Sci. Stat. Comp. 4 (1983) 261-269 .
5. VAN LOAN, C. On the method of weighting for equality constrained least squares problems, SIAM J. Numer. Anal. (1985) to appear.
6. VAN LOAN, C. The block Jacobi method for computing the singular value decomposition, Cornell Computer Science Technical Report TR 85-680, 1985.

Charles Van Loan
Department of Computer Science
Cornell University
Ithaca, New York 14853
USA

P S MARCUS

Numerical simulation of quasi-geostrophic flow using vortex and spectral methods

ABSTRACT

We present numerical solutions to the quasi-geostrophic equations that govern the flow of an incompressible fluid in a rapidly rotating annulus with a flat upper surface and a radially sloping bottom surface. These equations also govern the compressible flow in a rapidly rotating planetary atmosphere with an exponentially decreasing vertical density. We solve the equations as an initial-value problem and present pictures of the flow's vorticity as a function of time. We find that solving the flow with vortex methods leads to spurious results at late times due to the fact that high moments of the vorticity (enstrophy) are conserved poorly. Spectral and de-aliased spectral methods are competitive in time with and provide higher resolution than the vortex methods. The spectral methods also treat the enstrophy in more physically realistic manner.

1. EQUATIONS

The equations that govern the quasi-geostrophic flow of a rotating planetary atmosphere are (to first approximation) the same as those for a fluid with constant density ρ in a rotating annulus with inner radius a, outer radius b and where the bottom of the annulus has a constant slope s such that the depth d(r) of the bottom surface is

$$d(r) = sr \tag{1.1}$$

The quasi-geostrophic approximation is easily derived from the Euler equation by simultaneously expanding in two small dimensionless parameters: α the aspect ratio (ratio of depth of the annulus to the gap size (b-a)), and the Rossby number Ro (which is the ratio of the strength of the inertial force to the Coriolis force). The resulting approximation yields governing equations for a divergence-free, two-dimensional flow. (The flow along the axis of rotation is suppressed by the Taylor-Proudman theorem and the vertical boundaries). We obtain a non-dimensionalized equation for the vorticity ω along the axis of rotation as observed in a frame rotating with angular velocity Ω (see Pedlosky, 1979 Chapter 3)

$$[\frac{\partial}{\partial t} + (V_{\perp}\cdot\nabla)](\omega - F\psi + \beta r) = \frac{D_{\perp}}{Dt}(\omega - F\psi + \beta r) = 0 \tag{1.2}$$

where $V_{\perp}$ is the horizontal component of the velocity, ψ is the stream function (such that $V_{\perp} = \hat{e}_z \times \nabla\psi$ where $\hat{e}_z$ is the unit vector along the axis of rotation), and β is the non-dimensionalized slope

$$\beta = 2\Omega(b-a)s/\alpha U \tag{1.3}$$

where U is the characteristic horizontal velocity. In equation (1.2) the units of time, mass, and length are $(b-a)/U$, $\rho(b-a)^3$, and $(b-a)$. For an annulus with a free upper surface, the dimensionless parameter F is defined as

$$F = 4\Omega^2(b-a)/\alpha g \tag{1.4}$$

where g is the acceleration due to gravity. For an annulus covered by a lid, F is identically equal to zero. (In a rotating planetary atmosphere F is a measure of the vertical stratification.) The radius ratio $\eta = a/b$ appears in the boundary conditions since we require that the normal component of the velocity vanish at the radial boundaries of the annulus

$$V_r(|\underline{r}| = \frac{\eta}{1-\eta}, t) = V_r(|\underline{r}| = \frac{1}{1-\eta}, t) = 0 \tag{1.5}$$

Incompressibility is maintained by imposing

$$\nabla\cdot V_{\perp} = 0 \tag{1.6}$$

Note that $V_{\perp}$ is not a function of depth z.

If viscosity is included in the flow then the equations are modified by adding the term

$$Re^{-1}\nabla^2\omega \tag{1.7}$$

to the right-hand side of equation (1.2) where Re is the Reynolds number, Re $= U(b-a)/\nu$, and ν is the kinematic viscosity. However, further terms must be added to equation (1.2) that represent the additional viscous losses of energy and momentum from the interior flow by the Ekman pumping in the top and bottom viscous boundary layers. In this paper we consider only viscous flows for the case where the upper surface of the annulus is bounded by a lid (i.e. $F = 0$). The two-dimensional flow far from the boundaries is then approximated by (see Pedlosky 1979, Chapter 4):

$$\frac{D_\perp}{Dt}(\omega + \beta r) = [-\alpha^{-1}(\mathrm{ReRo})^{\frac{1}{2}} + \mathrm{Re}\,\nabla^2]\omega \tag{1.8}$$

where $\mathrm{Ro} = U/(2(b-a)\Omega)$ is the Rossby number. Equation (1.8) can be integrated to give an equation for $V_\perp$

$$\begin{aligned}\frac{D_\perp V_\perp}{Dt} = &-\beta r\hat{e}_z \times V_\perp + \mathrm{Re}^{-1}\nabla^2 V_\perp \\ &-\nabla\Pi - \alpha^{-1}(\mathrm{ReRo})^{\frac{1}{2}} V_\perp\end{aligned} \tag{1.9}$$

where Π is the pressure head.

2. VORTEX METHODS

Vortex methods using points, finite cores, and blobs (Chorin 1973, Anderson and Greengard 1984) and vortex-in-cell methods (Aref and Siggia 1981) are well known techniques for studying incompressible inviscid two-dimensional flow. Recently these techniques have been modified for viscous (Leonard 1980) and three-dimensional flows (Beale and Majda 1984) as well. The primary advantage of vortex methods is that they are gridless Lagrangian techniques where the computational elements automatically advect to regions of interest (i.e., where there is vorticity) and are not used where the flow is featureless (i.e. where the flow is irrotational or has uniform vorticity). The method of vortices works with inviscid, two-dimensional flow because the curl of Euler's equation

$$\frac{D_\perp \omega}{Dt} = 0 \tag{2.1}$$

shows that the vorticity ω is a Lagrangian invariant; that is, an element (point, core, or blob) of vorticity is advected by the velocity such that its strength is unchanged. If, for example, the vorticity is represented by a sum of delta function elements or point vortices

$$\omega(\underline{r},t) = \sum_i a_i \delta(\underline{r} - \underline{r}_i(t)) \tag{2.2}$$

The strength a_i of each element is constant in time and the guiding centers of each element advected with the flow

$$\frac{D_\perp}{Dt} \underline{r}_i(t) = V_\perp(\underline{r}_i(t),t) . \tag{2.3}$$

The velocity is found by using the Biot-Savart law with appropriate boundary conditions. In the point vortex example, an annulus with no central cylinder ($\eta = 0$) has velocity

$$V_\perp(\underline{r},t) = \hat{e}_z \times \sum_i \frac{a_i}{2\pi} \nabla \ell n(|\underline{r} - r_i(t)|) + V_{img} \tag{2.4}$$

where V_{img} is the velocity due to image vortices exterior to the annulus which insure that the normal component of the velocity at the boundary is zero

$$V_{img} = -\hat{e}_z \times \sum_i \frac{a_i}{2\pi} \nabla \ell n(|\underline{r} - \underline{r}_i'(t)|) \tag{2.5}$$

and where the image vortices are located at $\underline{r}_i'(t)$

$$\underline{r}_i' = (1-\eta)^{-2} \underline{r}_i / (\underline{r}_i \cdot \underline{r}_i) . \tag{2.6}$$

The images of cores and blobs are more complicated, and if $\eta \neq 0$, the images form an infinite series. Furthermore when $\eta \neq 0$ the right-hand side of equation (2.4) contains an additional term of the form $c/r\ \hat{e}_\phi$ that represents the irrotational flow due to vorticity at the inner boundary. With $\eta \neq 0$, the vortex-in-cell method is easier to use, since the boundary conditions are imposed when the Laplacian is inverted to find ψ from ω.

At first glance, it appears that vortex methods cannot be used for the flow in an annulus with a sloping bottom because equation (1.2) shows that the vorticity is not a Lagrangian invariant. If the vorticity is represented by discrete elements as in equation (2.2) then the strengths a_i

are no longer constant in time, and equation (2.3) for the guiding centers is no longer valid. However, we can still use vortex methods by utilizing the fact that the potential vorticity

$$\omega' = \omega - F\psi + \beta r \tag{2.7}$$

is a Lagrangian invariant. In this case we write the potential vorticity as a discrete sum of elements

$$\omega'(\underline{r},t) = \sum_i a_i \delta(\underline{r} - \underline{r}_i(t)) \tag{2.8}$$

For point vortex elements the total velocity is now equal to

$$V_\perp(\underline{r},t) = \begin{cases} \hat{e}_z \times \sum_i a_i (2\pi)^{-1} \nabla \ln(|\underline{r} - \underline{r}_i(t)|) & F = 0 \\ -\hat{e}_z \times \sum_i a_i (2\pi)^{-1} \nabla K_0(F^{\frac{1}{2}}|\underline{r} - r_i(t)|) & F \neq 0 \end{cases}$$

$$+ V_{img} + V_{bac}(\underline{r},t)\hat{e}_\phi \tag{2.9}$$

where

$$V_{bac}(\underline{r},t) = \begin{cases} (c/r - Br^3/3)\hat{e}_\phi & F = 0 \\ -\beta \int dr' r' K_0[r - r')] & F \neq 0 \end{cases} \tag{2.10}$$

where V_{img} is the velocity due to the image vortices and where $K_0(r)$ is a modified Bessel function of the second kind of order zero. The strengths a_i are constant in time, and the guiding centers of the elements obey equation (2.3).

Note that with the Euler equation and the usual method of vortex elements used in equation (2.4), the condition $V_\perp=0$ corresponds to no vortex elements present. If a flow consists of no motion except for one or two isolated spots or sheets of vorticity then very few vortex elements are needed, and the method is economical. This economy does not seem to be present for the complicated geometry of the annulus with the sloping bottom surface. Equation (2.9) shows that when no elements are present, the velocity is not equal to zero. Instead the velocity is the azimuthal flow V_{bac} given by equation (2.10). Representing the flow $V_{\perp}=0$ requires a large number of elements. For example, if F=0 then $V_{\perp}=0$ corresponds to an axially symmetric distribution of vortices with a density that increases linearly with radius. Therefore these vortex methods are not economical for flows with a few isolated sheets or spots of vorticity.

However, these vortex methods are useful for the types of flows we wish to study in this paper. The motivation for this work was to develop a numerical method for the study of the flow of the Great Red Spot of Jupiter and the azimuthal zonal flow in which it is located. Curiously enough V_{bac}, the azimuthal flow of equation (2.10) is a good approximation to the zonal flow near the Red Spot (Ingersoll et al. 1979). Therefore these vortex methods can be used efficiently. The vortex elements represent the vorticity of the Red Spot and the small deviation of the zonal flow from V_{bac}.

These vortex methods are also useful for planned laboratory experiments (by Harry L. Swinney at the University of Texas Austin) in which a fluid will be rotated rapidly in an annulus with a fixed upper lid and a sloping bottom. In these experiments the fluid will be stirred randomly and strongly. In general, mixing causes Lagrangian invariant quantities to become (macroscopically) uniform throughout the flow. (For example, a well stirred Boussinesq fluid for which the temperature is a Lagrangian invariant becomes macroscopically isothermal everywhere except at the boundaries.) The stirring should produce a uniform potential vorticity and a macroscopic velocity not too different from V_{bac}. We therefore expect the stirred fluid in an annulus with sloping bottom to have a flow that can be efficiently simulated by vortex methods.

3. SPECTRAL METHODS

A spectral method is an alternative way of solving equation (1.9). We have employed a de-aliased spectral code that we used previously to simulate the three-dimensional Taylor-Couette flow between two concentrically rotating cylinders. The numerical method is described in detail by Marcus (1984). With this method the velocity is expanded in Chebyshev modes in the radial and Fourier modes in the azimuthal directions.

$$V_\perp(\underline{r},t) = \sum_{k,m} V(k,m,t)e^{im\phi}T_k(2r - \frac{1+\eta}{1-\eta}) \tag{3.1}$$

A similar expansion is used for the pressure head. When expansion (3.1) is substituted into equation (1.9) we obtain a set of nonlinearly coupled ordinary differential equations in time for $V(k,m,t)$. The differential equations are solved by the method of fractional steps. In the first step the velocity is updated with the nonlinear advection terms, and in this step neither incompressibility nor boundary conditions are imposed.

$$V_\perp^{(1)}(t+\Delta t) = V_\perp(t-\Delta t) + 2\Delta t[\omega(t) - \beta r][V_\perp(t) \times \hat{e}_z] \tag{3.2}$$

Note that we have used the rotation form of equation(1.9). In the second fractional step the pressure head is computed such that when the gradient of the pressure is subtracted from $V^{(1)}(t+\Delta t)$ the resulting velocity is divergence-free.

$$V_\perp^{(2)}(t + \Delta t) = V_\perp^{(1)}(t + \Delta t) - \nabla\Pi \tag{3.3}$$

with

$$\nabla^2\Pi = \nabla\cdot V_\perp^{(1)}(t + \Delta t) \tag{3.4}$$

The Poisson operator in equation (3.4) is inverted using the Haidvogel-Zang factorization (1979), and the boundary conditions for the inversion are <u>implicitly</u> imposed by using the Greens′ function method discussed by Marcus such that the velocity at the end of all three fractional steps is divergence free and satisfies the correct <u>Cauchy</u> conditions at the radial boundaries. The third and last fractional step imposes the viscosity with a backwards Euler method.

$$V_\perp^{(3)}(t+\Delta t) = V_\perp^{(2)}(t+\Delta t) + [Re^{-1}\nabla^2 - (ReRo)^{\frac{1}{2}}]V_\perp^{(2)}(t+\Delta t) \tag{3.5}$$

An aliasing error is introduced because the nonlinear terms are not computed as a spectral convolution, but instead by collocation: the velocity is fast-transformed onto a grid, the nonlinear product is computed at the grid points, and then it is fast-transformed back to find the Chebyshev-Fourier coefficients of the product. This last transform will in general introduce an aliasing error. However it is easy to show (Gottlieb and Orszag 1977) that for quadratic nonlinearities if the grid size in each spatial dimension is 3/2 (or more) times greater than the number of spectral modes in the expansion, then the product is free of aliasing errors. We have used "the 3/2 rule" in our code.

4. RESULTS

We used vortex methods with points, finite cores, and blobs for geometries with $\eta=0$ and flows with no viscosity. We also used vortex-in-cell methods for viscous and inviscid flows with geometries with the inner radial boundary present and not present. All of our inviscid codes conserved energy and angular momentum to a high degree of accuracy. However, none of the vortex methods that we tried conserved enstrophy (the square of the potential vorticity integrated over the annulus) well. We found that in only three or four revolutions of the annulus the enstrophy decayed by a factor of two. The lack of conservation of enstrophy is consistent with the results found by Aref and Siggia with their vortex-in-cell calculations. Unfortunately, our goal in this research was to show how a long-lived Red Spot-like feature of vorticity forms. With the vortex methods we found that a spot-like feature began to form, but after several revolutions it would always decay due to the numerical dissipation of enstrophy.

Fortunately the inviscid spectral code not only conserved energy and angular momentum but also enstrophy to high accuracy. In figures 1 to 8 we have plotted the potential vorticity for a flow with $\eta = 0.25$, Re=5500., Ro = 0.01, and F = 0. Initial conditions were chosen with one spot of clockwise and one spot of counter-clockwise vorticity. Each spot is of nearly uniform strength, and the magnitude of each spot is approximately equal to the magnitude of the vorticity of V_{bac}. The figures show the potential vorticity plotted with a gray scale so that the whitest regions are rotating in the same sense as V_{bac} and the dark regions in the opposite sense. The number of rotations of the annulus is given in the figure captions. The pictures clearly show that the dark spot breaks up and produces macroscopically uniform filamentary structures whereas the white spot is stable and long lived.

In a second experiment shown in figures 9 - 16, we start with two white spots of potential vorticity whose strengths are about one third of the vorticity in V_{bac}. The spots quickly merge to form one spot. A third experiment which starts with an axisymmetric thick ring of potential vorticity is shown in figures 17 - 24. The m=3 eigenmode is the most linearly unstable mode. The eigenmode grows and forms three spots which then merge together.

I thank Mr. Nicholas Socci for producing the figures. I acknowledge the support of National Science Foundation grants AST-84-12170 and MEA-84-10412. The numerical calculations were performed at the National Center for Atmospheric Research which is sponsored by the NSF.

REFERENCES

Aref, H. and Siggia, E.D., J. F. M. 109 (1981) 435-463.

Anderson, C.R. and Greengard, C., Lawrence Berkeley Lab. Report (1984) lbl-16376.

Beale J.T. and Majda A., J. C. P. 58 (1985) 188-205.

Chorin A.J., J. F. M. 57 (1973) 785-803.

Gottlieb, D. and Orszag, S.A., Numerical Analysis of Spectral Methods, S.I.A.M., Phila., 1977.

Haidvogel, D.B. and Zang, T., J. C. P. 30 (1979) 167-180.

Ingersoll A.P., Beebe, R.F., Collins, S.A., Hunt, G.E., Mitchell, J.L., Muller, P., Smith, B.A., and Terrile, R.J., Nature 280 (1979) 773-775.

Leonard A., (1980) J. C. P. 37 (1980) 289-335.

Marcus, P.S., J. F. M.146 (1984) 45-64.

Pedlosky, J., Geophysical Fluid Dynamics Springer-Verlag, N.Y., 1979.

P.S. Marcus
Center for Astrophysics
Harvard University
60 Garden Street
Cambridge
Massachusetts
MA 02138
USA

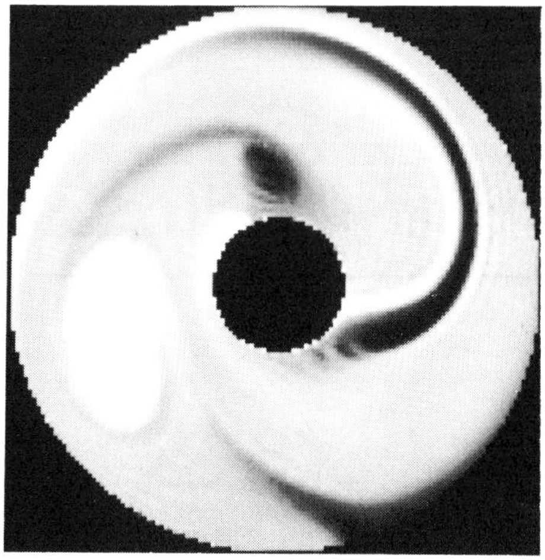

Figure 3 - 0.6 rotations

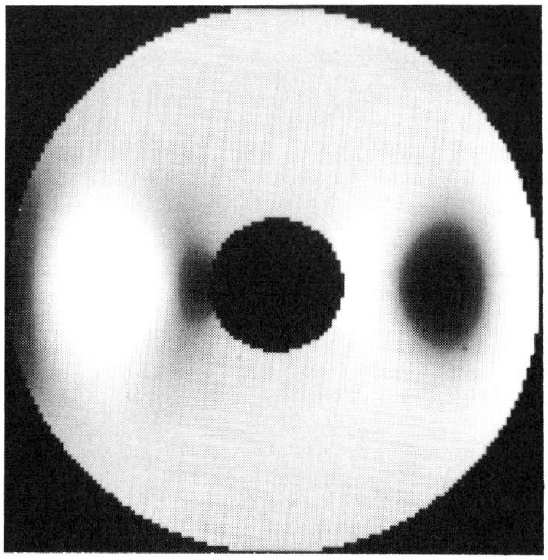

Figure 1 - 0 rotations

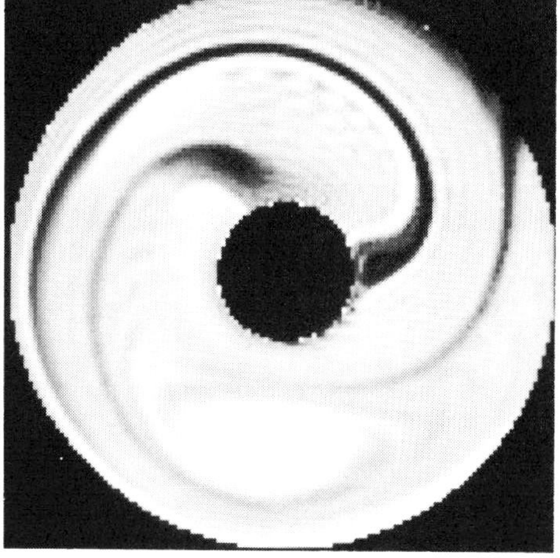

Figure 4 - 1.0 rotation

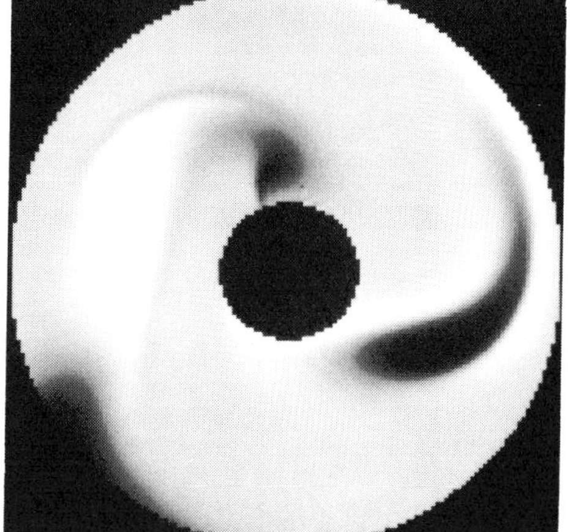

Figure 2 - 0.25 rotations

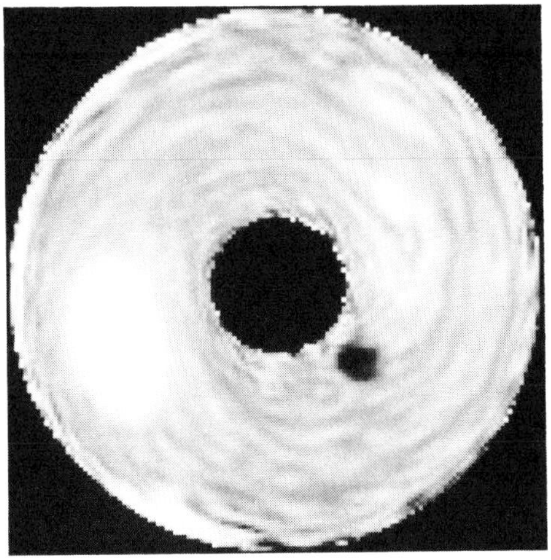

Figure 7 - 8.0 rotations

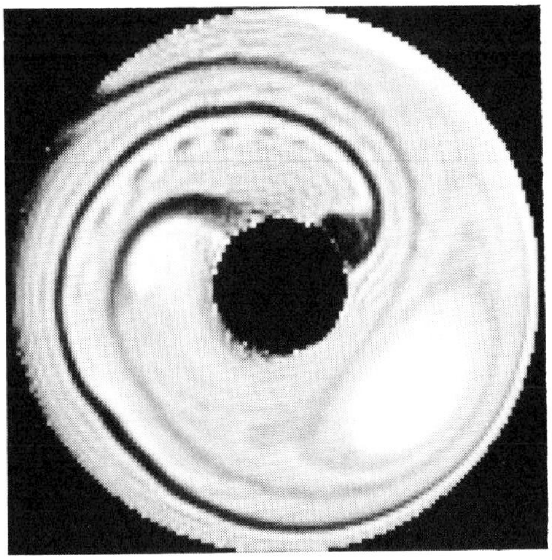

Figure 5 - 1.3 rotations

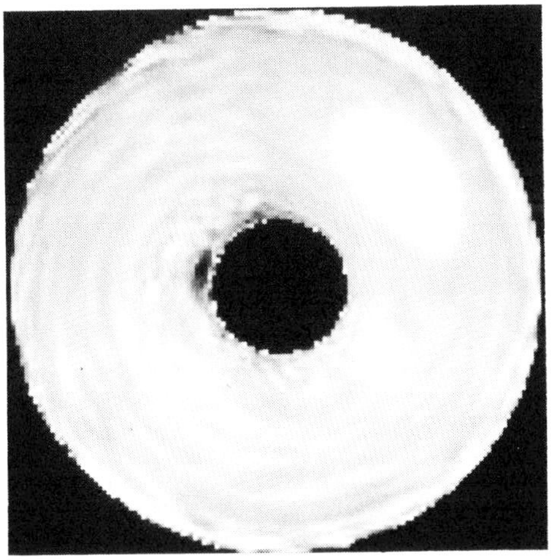

Figure 8 - 32.0 rotations

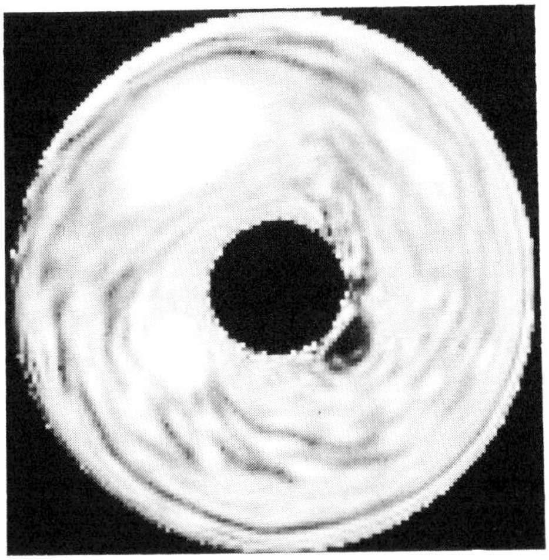

Figure 6 - 2.6 rotations

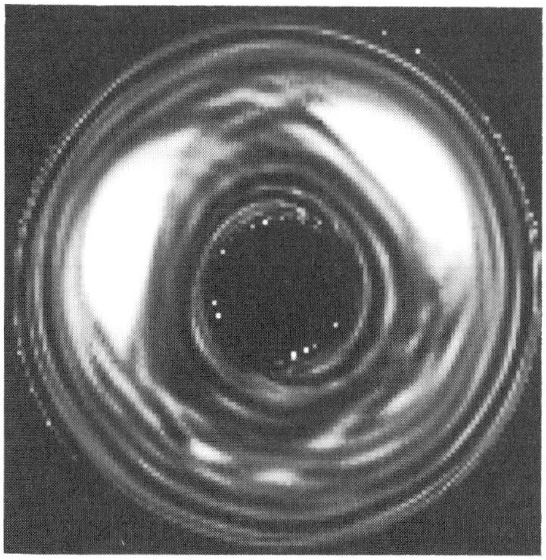

Figure 11 - 4.0 rotations

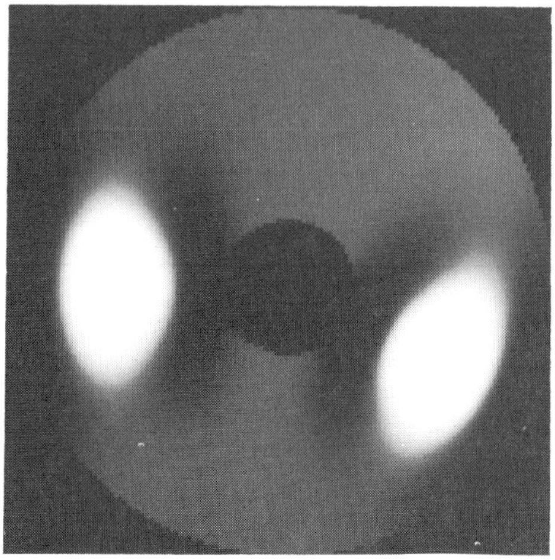

Figure 9 - 0 rotations

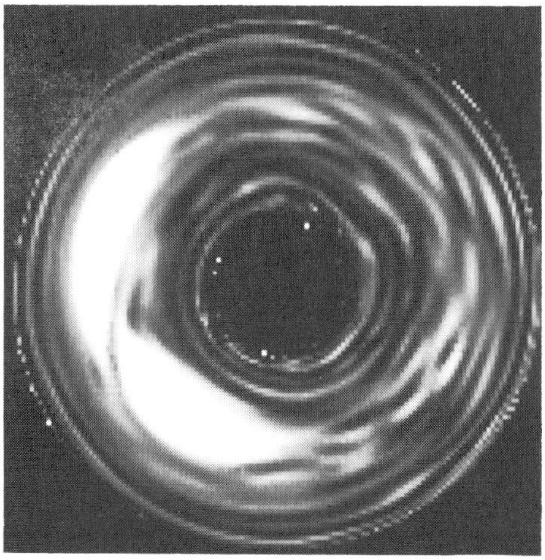

Figure 12 - 5.3 rotations

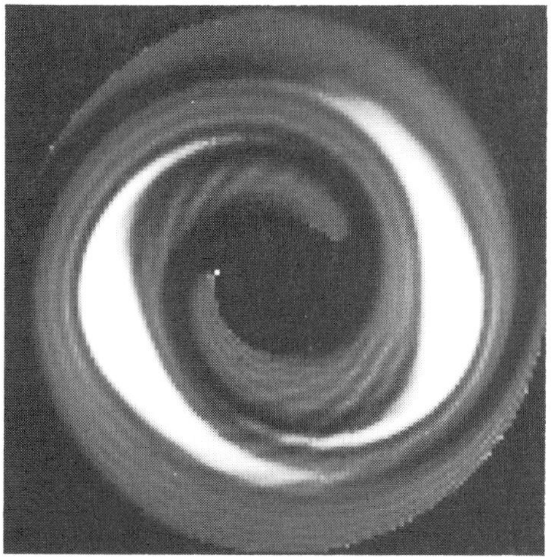

Figure 10 - 1.0 rotation

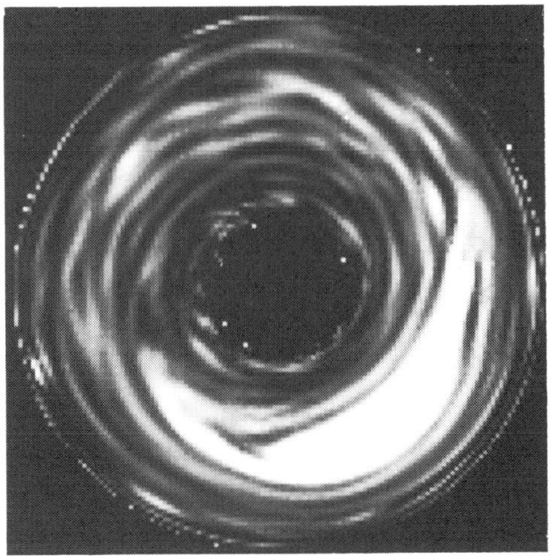

Figure 15 - 9.3 rotations

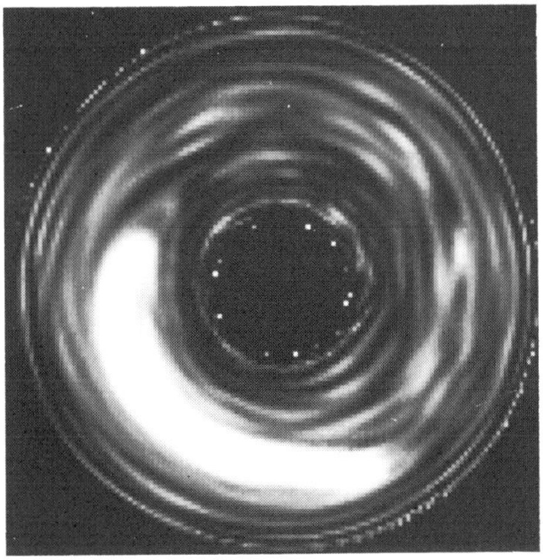

Figure 13 - 6.6 rotations

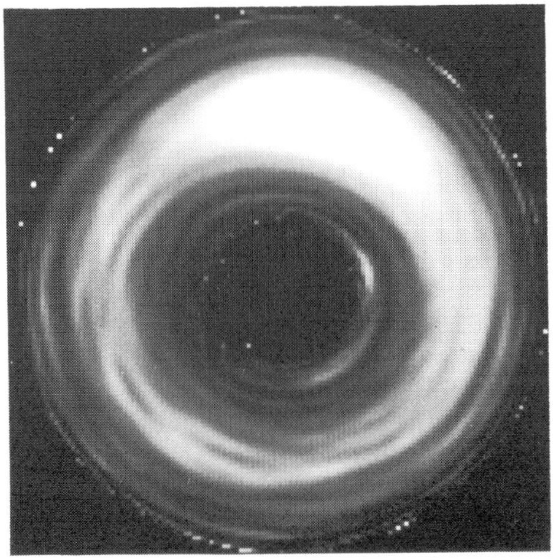

Figure 16 - 33.3 rotations

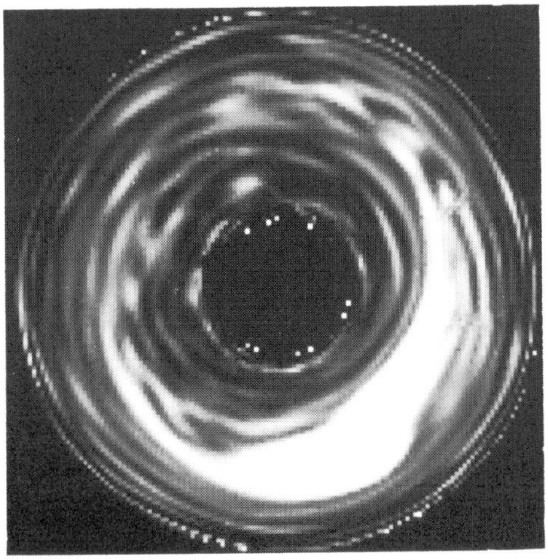

Figure 14 - 8.0 rotations

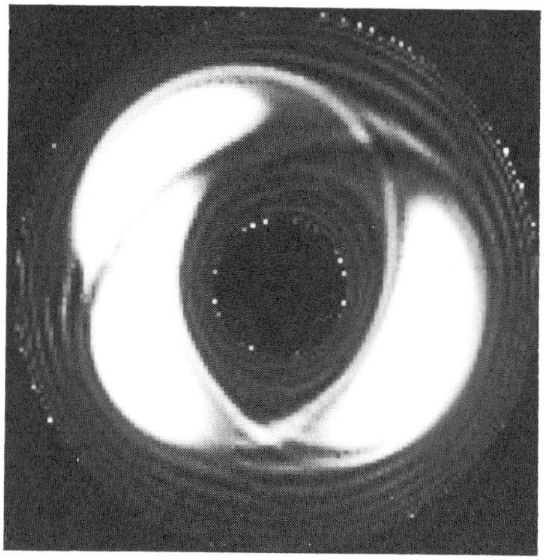

Figure 19 - 2.7 rotations

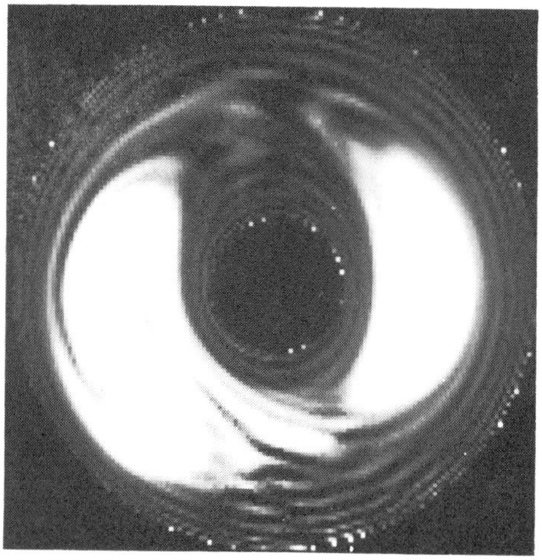

Figure 20 - 3.1 rotations

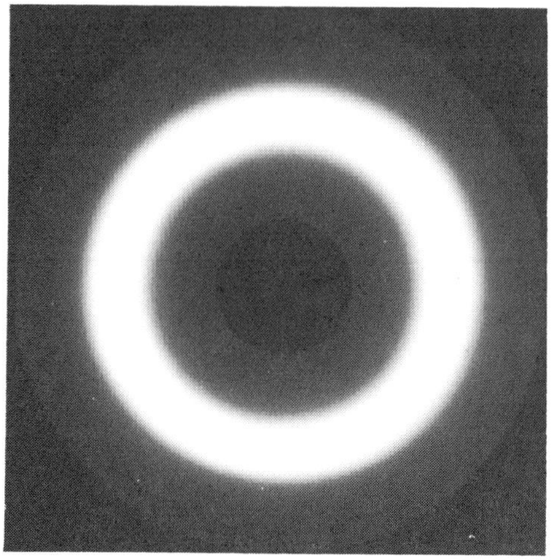

Figure 17 - 0 rotations

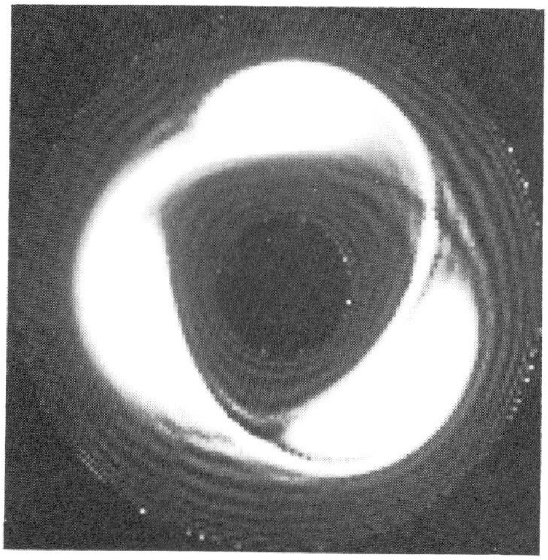

Figure 18 - 2.3 rotations

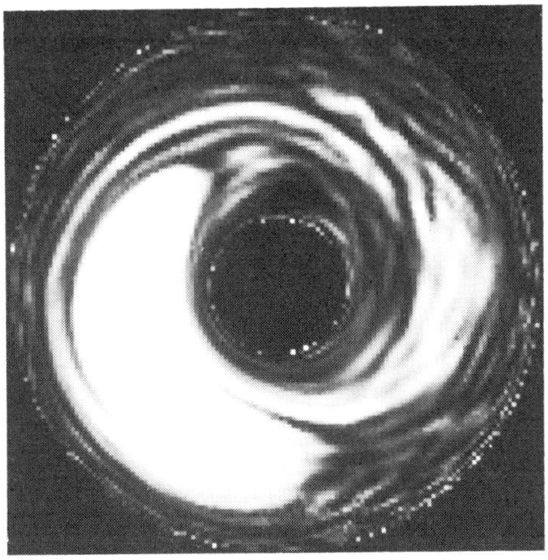

Figure 23 - 5.6 rotations

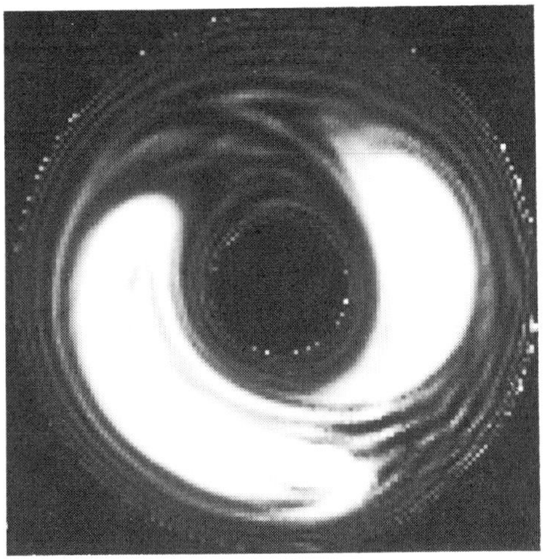

Figure 21 - 3.5 rotations

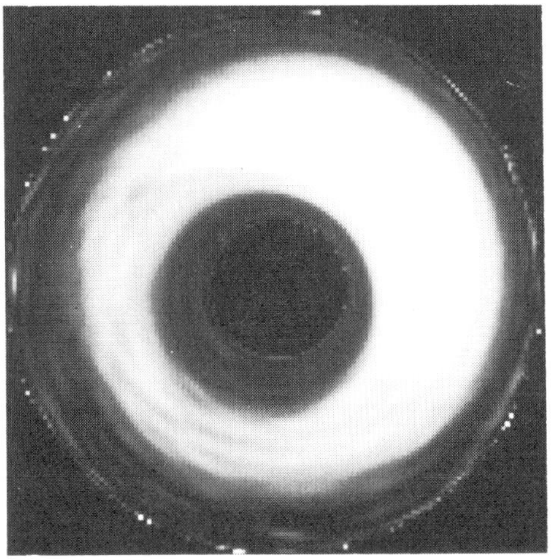

Figure 24 - 20.5 rotations

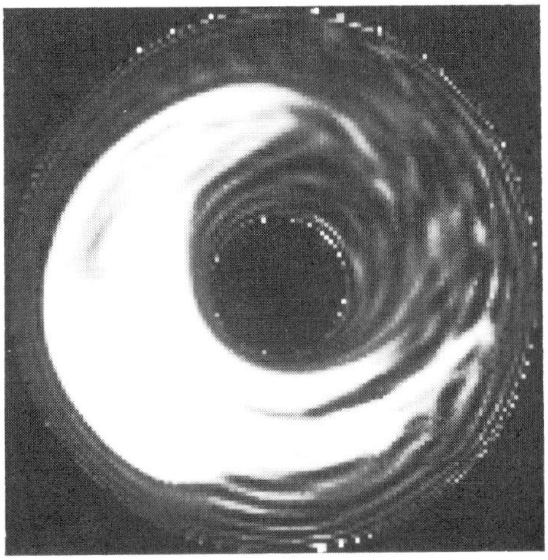

Figure 22 - 4.0 rotations

A R MITCHELL & D F GRIFFITHS

Beyond the linearised stability limit in nonlinear problems

1. INTRODUCTION

In recent years a new type of "numerical analyst" has emerged. This is a person probably from physics or biology, equipped with a small computer, who carries out long time calculations with a fixed time step of non linear problems, the latter modelled either by differential or difference equations. Not for this person is a well behaved solution, unique at each point in space and time being sought, but more likely phenomena such as bifurcation, period doubling, chaos, recurrence etc. Most desirable of these is chaos where the solutions have stochastic behaviour even although the governing equations are deterministic. This has given fluid dynamicists a new hope of understanding the onset of turbulence in a viscous flow where the Reynolds Number parameter is being increased.

Non linear problems tend to be individual and in this paper we shall concentrate on two different examples of non linear instability and its effect on the respective solutions. The examples are taken from

(I) Mathematical Biology and the non linear terms $u(1-u)$.
(II) Dispersive waves and the Korteweg de Vries equation.

2. THE NON LINEAR TERM $u(1-u)$

2.1 Map on the line

In population dynamics, May [1] considered the difference equation model (or map on the line)

$$V_{n+1} = aV_n(1-V_n) , \quad n = 0,1,2,... \tag{2.1}$$

where $a(>0)$ is a parameter. If we put

$$V_n = \frac{\alpha k}{1+\alpha k} U_n \quad \text{and} \quad a = 1 + \alpha k$$

into (2.1), we get

$$U_{n+1} = (1+\alpha k)U_n - \alpha k U_n^2 \tag{2.2}$$

which is the forward Euler difference replacement of the equation

$$\frac{du}{dt} = \alpha u(1-u) , \qquad \alpha > 0 \tag{2.3}$$

where $\alpha(>0)$ is a parameter and $k(>0)$ is the time step. The solution $u = 1$ is a stable rest point of (2.3) and if we put

$$U_n = 1 + \varepsilon_n$$

in (2.2), where ε_n is small, and linearise, $U = 1$ emerges as a stable rest point of (2.2) provided

$$0 < \alpha k \leq 2 . \tag{2.4}$$

Returning to May's notation, it follows that linearised stability is present about the asymptotic solution

$$V = \frac{a - 1}{a} \tag{2.5}$$

provided

$$1 < a \leq 3 . \tag{2.6}$$

A series of numerical experiments using (2.1) was carried out for a range of values of the parameter a, and the results shown in Figure 1. In each experiment, the initial condition was taken to be $U_0 = \frac{1}{2}$ and the iteration continued until the asymptotic (in time) solution was obtained. Period doubling started at $a = 3$ and continued until $a \doteqdot 3.57$ when chaos was encountered. If period 2^j, j an integer, occurs for the parameter a_j and $a_\infty = \lim_{j\to\infty} a_j$, then

$$\frac{a_\infty - a_j}{a_\infty - a_{j+1}} \to 4.6692 \quad \text{as} \quad j \to \infty ,$$

where 4.6692 is the Feigenbaum constant [2].

2.2 Map on the plane

The quadratic map introduced first by Hénon [3] is

$$\begin{aligned} x_{n+1} &= 1 - \mu x_n^2 + y_n , \\ y_{n+1} &= b x_n , \end{aligned} \qquad n = 0,1,2,\ldots \tag{2.7}$$

where μ and b are parameters. If we put

$$x = \frac{1}{\mu}[az - \tfrac{1}{2}(a+1-b)] \quad \text{and} \quad \mu = \tfrac{1}{4}[a^2 - (1-b)^2] \tag{2.8}$$

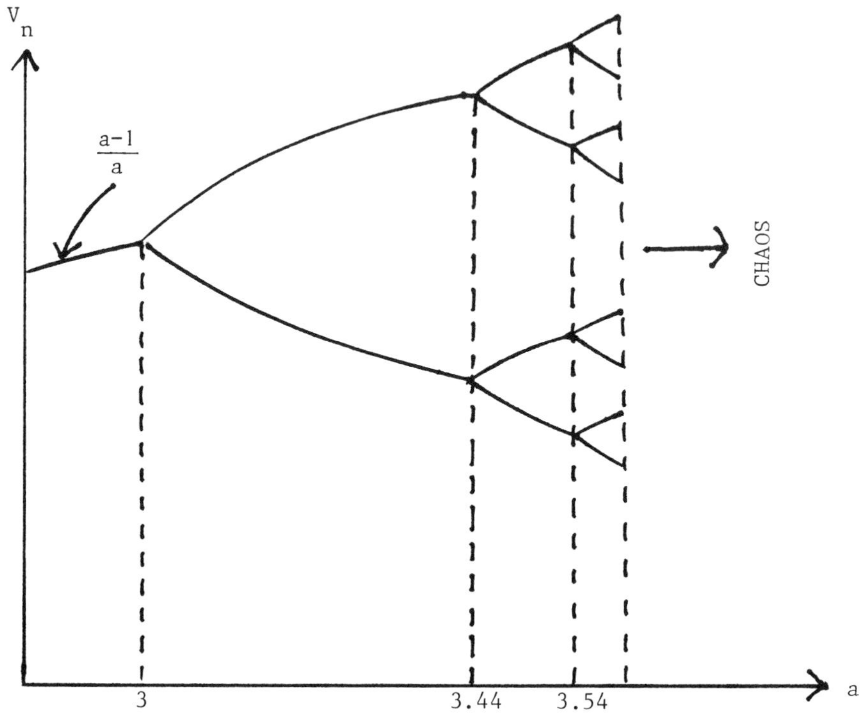

Figure 1. Period doubling leading to chaos

(2.7) reduces to

$$z_{n+1} = az_n(1-z_n) + (1-b)z_n + bz_{n-1}, \quad n = 1,2,3,\ldots \ . \tag{2.9}$$

If $a = 2\alpha k$ and $b = 1$, (2.9) reduces to

$$z_{n+1} = 2\alpha k z_n(1-z_n) + z_{n-1} \tag{2.10}$$

which is the leap frog (or mid point) difference replacement of (2.3). In the phase plane, z_{n+1} plotted against z_n, there are two unstable fixed points (0,0) and (1,1) and two period 2 points (1,0) and (0,1). Following Sanz-Serna [4] we rewrite (2.10) as the augmented system

$$\begin{aligned} u_{n+1} &= u_n + 2\alpha k v_n(1-v_n) \\ v_{n+1} &= v_n + 2\alpha k u_{n+1}(1-u_{n+1}) \end{aligned} \tag{2.11}$$

where $u_n = z_{2n}$ and $v_n = z_{2n+1}$, $n = 0,1,2,\ldots$. The system (2.11) in the phase plane, v_n plotted against u_n, has two unstable fixed points (0,0) and (1,1) and two neutrally stable fixed points (1,0) and (0,1) provided $\alpha k \le 1$.

The qualitative behaviour of solutions of (2.11) and consequently (2.10) for small time steps may be deduced by observing that (2.11) is consistent with the augmented system

$$\frac{du}{dt} = \alpha v(1-v)$$

$$\frac{dv}{dt} = \alpha u(1-u)$$

which contains the solutions of (2.3) as the special case $u = v$. This system has the first integral

$$\tfrac{1}{2}u^2 - \tfrac{1}{3}u^3 = \tfrac{1}{2}v^2 - \tfrac{1}{3}v^3 + \text{constant}$$

from which it follows that the invariant curves are given by the diagonal $u = v$ and the ellipse

$$(u + v - 1)^2 + \tfrac{1}{3}(u-v)^2 = 1.$$

With initial values chosen inside this ellipse, the solutions describe orbits around the centres (0,1) and (1,0) while points outside the ellipse "blow up" along $u \sim v \to -\infty$ in finite time. These qualitative properties of the augmented system clearly describe the behaviour of the trajectories of (2.11) shown in Figure 2 for $\alpha k = 0.1$, and a variety of initial conditions. This has previously been noted by Sanz-Serna [4].

The solutions of (2.11) depend on the three parameters u_0, v_0 and αk. However, since (2.11) is, through (2.10), a difference replacement of (2.3), then u_0 and v_0 should not be treated as independent parameters but should be chosen so that the numerical solution follows the unstable manifold joining (0,0) to (1.1). For finite values of αk, this unstable manifold denoted by $v = \phi(u)$ is not linear but is smooth in the neighbourhood of the origin and can be expressed as the Taylor series

$$\phi(u) = a_1 u + a_2 u^2 + a_3 u^3 + \ldots$$

where $a_1 = \alpha k + (1 + \alpha^2 k^2)^{\frac{1}{2}}$,

$$a_2 = 2\alpha k a_1^2/(2\alpha k - 1 - a_1 - 2\alpha k a_1)$$

and $a_3 = 2a_1 a_2(2\alpha k + a_2)/(2\alpha k - a_1 - a_1^3)$.

As the unstable manifold approaches the saddle point (1,1) it oscillates with increasing frequency and amplitude and becomes wrapped around the orbital solutions neighbouring (1,0) and (0,1) while being bounded externally by the two unstable manifolds that extend from (1,1) to the origin. (This is most clearly seen in Figure 4 for $\alpha k = 0.5$.) All such solutions eventually escape to infinity along $u \sim v \to -\infty$.

The numerical simulations shown in Figures 3-5 for $\alpha k = 0.3$, 0.5, and 0.95 were obtained by choosing a wide variety of initial conditions in the range $10^{-11} \leq u_0 \leq 10^{-7}$ and computing the corresponding initial values for v_0 from the Taylor series $v_0 = \phi(u_0)$. The number of time steps executed before blow-up was encountered (measured by $|u_n| + |v_n| > 100$) varied erratically from approximately 25 to 5,000 depending on the number of cycles performed around the centres.

Another interesting feature of the solutions of (2.11) is the transition from orbital solutions around the centres (1,0) and (0,1). The insets in Figures 4 and 5 show the appearance of islands (obtained by judicious choices of initial values (u_0, v_0)); during each cycle around the centre, one point is placed on each island. The 'centres' of these islands therefore constitute periodic solutions of (2.11), (period 7 for $\alpha k = 0.5$ and period 8 for $\alpha k = 0.95$) while the immediate exterior of the islands is part of the unstable manifold emanating from the origin.

As αk is increased beyond unity, the fixed points (0,1) and (1,0) are no longer centres and consequently orbits do not exist although there may still be periodic solutions.

2.3 Chaos via Hopf Bifurcation (Map in 4-space)

An alternate route to chaos, via Hopf bifurcation, is provided by the Adams predictor-corrector scheme

$$\begin{aligned} U_{n+1}^{(p)} &= U_n + \frac{k}{24}[55U_n(1-U_n) - 59U_{n-1}(1-U_{n-1}) + 37U_{n-2}(1-U_{n-2}) \\ &\qquad - 9U_{n-3}(1-U_{n-3})] \\ U_{n+1} &= U_n + \frac{k}{24}[9U_n^{(p)}(1-U_n^{(p)}) + 19U_n(1-U_n) - 5U_{n-1}(1-U_{n-1}) \\ &\qquad + U_{n-2}(1-U_{n-2})] \end{aligned} \tag{2.12}$$

where we have put $\alpha = 1$. Linearisation about $U = 1$ in the usual way produces the quartic equation

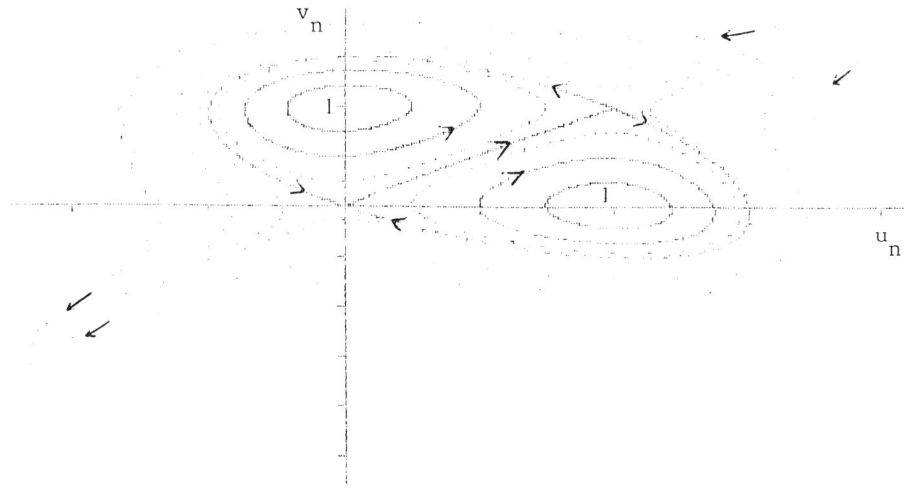

Figure 2 - Solutions of (2.11) for $\alpha k = 0.1$ from a variety of initial points (u_0, v_0).

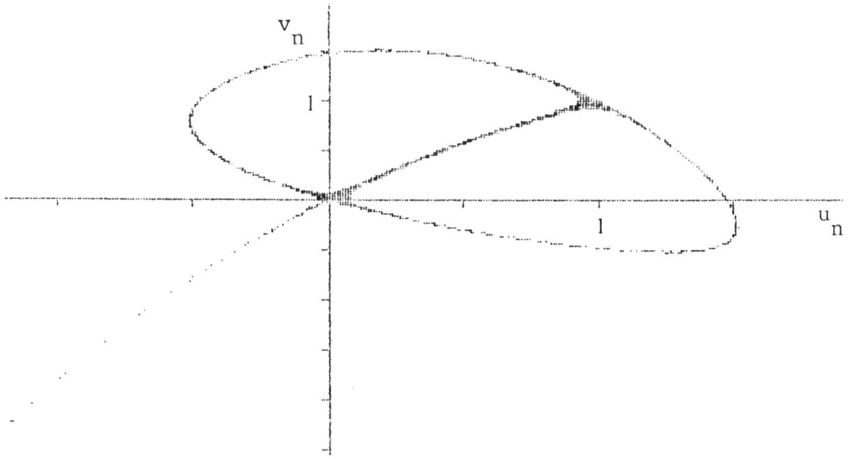

Figure 3 - Solutions of (2.11) for $\alpha k = 0.3$ and initial values $(u_0, \phi(u_0))$, $0 < u_0 \ll 1$.

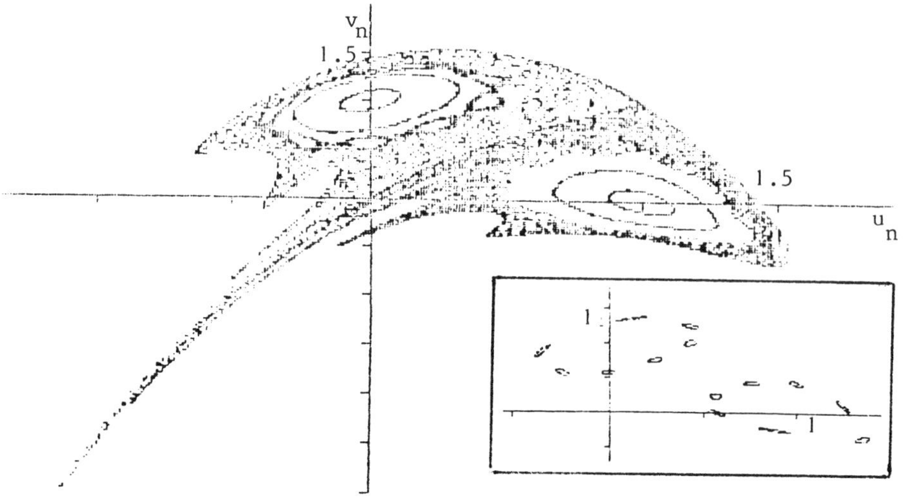

Figure 4 - Solutions of (2.11) for $\alpha k = 0.5$ with initial values i) $(u_0, \phi(u_0)$ $(0 < u_0 \ll 1)$ and ii) chosen to produce orbital solutions. The inset (reduced) shows the islands which exist in the transition zone between the image of the unstable manifold and the region of orbital solutions.

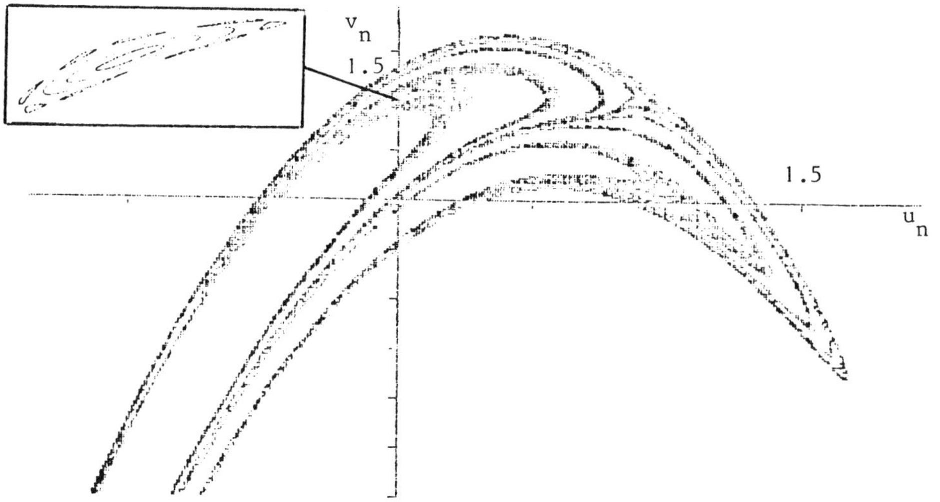

Figure 5 - Solutions of (2.11) for $\alpha k = 0.95$ and initial values $(u_0, \phi(u_0))$ $(0 < u_0 \ll 1)$. The inset shows the solutions in the neighbourhood of the centre (0,1).

$$\lambda^4 - (1 - \tfrac{7}{6}k + \tfrac{55}{64}k^2)\lambda^3 - (\tfrac{5}{24}k - \tfrac{59}{64}k^2)\lambda^2 + (\tfrac{1}{24}k - \tfrac{37}{64}k^2)\lambda + \tfrac{9}{64}k^2 = 0 \qquad (2.13)$$

for the amplification factor λ. The locations of the roots λ for the following values of k are as follows

$k = 0 \qquad \lambda = 1,0,0,0$

$k \doteqdot 1.28 \qquad \lambda \approx \bullet,\bullet,+i,-i$

$k = 8/3 \qquad \lambda = 1,1,1,1,$

where $\bullet$ signifies that the root is inside the unit circle in the complex plane. For the range $0 < k \le \frac{8}{3}$ two of the roots are inside or on the unit circle whereas the other two roots are inside or on the unit circle for $0 < k < 1.28$ and outside the unit circle for $1.28 < k < 8/3$. Hopf bifurcation [5] occurs at $k \doteqdot 1.28$ and for $1.28 < k < 1.60$ (approximately) the attractors are invariant limit cycles. For $k > 1.60$, the limit cycles break down giving chaos via period doubling. Detailed calculations for $k > 1.28$ can be found in Prüfer [6].

2.4 Chaos in Reaction-Diffusion

We now study the effect on bifurcation of adding diffusion to (2.1) resulting in Fisher's equation with quadratic non-linearity

$$\frac{\partial u}{\partial t} = \alpha u(1-u) + \beta \frac{\partial^2 u}{\partial x^2}, \quad \alpha,\beta > 0 . \qquad (2.14)$$

The forward Euler difference replacement of (2.14) is

$$U_m^{n+1} = U_m^n + \alpha k U_m^n(1-U_m^n) + \frac{\beta k}{h^2}(U_{m+1}^n - 2U_m^n + U_{m-1}^n) \qquad (2.15)$$

where $x = mh$ and $t = nk$, with m and n integers. A standard Von Neumann linearised stability analysis of (2.15) leads to the condition

$$0 \le \alpha k(U_m^n - \tfrac{1}{2}) \le 1 - \frac{2\beta k}{h^2}, \qquad (2.16)$$

and so adding diffusion appears to lower the linearised stability threshold.

We now carry out some numerical experiments where we choose a bifurcation free solution based on (2.1) with $a < 3$ and add diffusion to the problem by considering $\beta > 0$. Reverting to May's notation, (2.15) becomes

$$V_m^{n+1} = aV_m^n(1-V_m^n) + b(V_{m+1}^n - 2V_m^n + V_{m-1}^n) \qquad (2.17)$$

where $a = 1 + \alpha k$ and $b = \frac{\beta k}{h^2}$. The problem investigated is (2.17) subject to the initial condition

$$V(x,0) = \begin{cases} 0.5 & x = 0 \\ 0 & 0 < x \le 100 \end{cases}$$

and the boundary conditions

$$\frac{\partial V(0,t)}{\partial x} = 0 \qquad \forall\ t > 0$$

$$V(L,t) = 0 \qquad \forall\ t > 0$$

where L is sufficiently large so as not to affect the results near $x = 0$. The numerical results obtained with $h = 0.5$, $k = 0.125$, $L = 100$ are given for $a = 2.915$ and increasing b as follows

b	V_0^n
$0 \le b \le 0.01$	0.65694683
$0.0125 \le b \le 0.1275$	Periodic
$0.1300 \le b \le 0.2150$	Chaos
$0.2150 < b$	Overflow

where for each value of b, up to 10,000 time steps were required to obtain the behaviour of the solution for large time. Further experimental corroboration of the destabilising effect of diffusion is available in Mitchell and Bruch [7].

3. NON-LINEAR DISPERSIVE WAVES

3.1 The Korteweg-de-Vriés (K.d.V.) equation

We consider next the KdV equation

$$\frac{\partial u}{\partial t} + u\frac{\partial u}{\partial x} + \varepsilon \frac{\partial^3 u}{\partial x^3} = 0 , \qquad \varepsilon > 0 . \tag{3.1}$$

For the linearised version of (3.1)

$$\frac{\partial u}{\partial t} + U \frac{\partial u}{\partial x} + \varepsilon \frac{\partial^3 u}{\partial x^3} = 0 , \qquad U \text{ constant,}$$

periodic wavetrain solutions are given by

$$u = ae^{i(kx-wt)} \tag{3.2}$$

where a is the amplitude, k the wave number, and w the frequency, and the dispersion relation is

$$w = k(U - \varepsilon k^2) .$$

For the nonlinear equation (3.1), however, (3.2) is not a solution and the dispersion relation, if it could be found, would involve the amplitude. Further details of dispersive waves can be found in the excellent treatise by Whitham [8].

3.2 Stability of the KdV equation

In order to examine the stability of the KdV equation we consider a small perturbation $\delta(x,t)$ on the solution $u(x,t)$ leading to the perturbation equation

$$\delta_t + u\delta_x + u_x\delta + (\tfrac{1}{2}\delta^2)_x + \varepsilon\delta_{xxx} = 0 . \tag{3.3}$$

Multiply (3.3) by δ and integrate with respect to x over a finite space interval I. For periodic boundary conditions, we obtain the result

$$\frac{d}{dt}\|\delta(t)\|^2_{L_2} + \int_I \delta^2(x,t)\,\frac{\partial u(x,t)}{\partial x}\,dx = 0 . \tag{3.4}$$

If $\frac{\partial u}{\partial x} < 0$, then $\frac{d}{dt}\|\delta(t)\|^2_{L_2} > 0$ and so $\delta(t)$ increases and it would appear that a negative gradient in the initial condition for example would trigger off an instability in the solution of (3.1). Thus although the zero dispersion limit for the KdV equation is complicated (Lax [9]) it would appear that this result is consistent with the appearance of a shock due to converging straight characteristics when $\varepsilon = 0$ and a negative gradient is present in the initial condition.

3.3 Discretisation in Space

Using equally spaced finite elements with piecewise linear basis functions the nonlinear term gives either

$$u\frac{\partial u}{\partial x} \sim \frac{1}{3}(U_{m+1} + U_m + U_{m-1})\frac{1}{2h}(U_{m+1} - U_{m-1}) \sim \frac{1}{2}\frac{\partial}{\partial x}(u^2)$$

for the standard Galerkin method where $u \sim U = \sum_i U_i\phi_i(x)$ or

$$\frac{1}{2}\frac{\partial}{\partial x}(u^2) \sim \frac{1}{2}(U_{m+1} + U_{m-1})\frac{1}{2h}(U_{m+1} - U_{m-1}) = \frac{1}{4h}(U^2_{m+1} - U^2_{m-1})$$

for product approximation [10] where $u^2 \sim U^2 = \sum_i U_i^2\phi_i(x)$. The mass matrix is formed from

$$u \sim U = \frac{1}{6}(U_{m+1} + 4U_m + U_{m-1}) = (1 + \frac{1}{6}\delta^2)U_m$$

where δ is the central difference operator.

With finite differences, the nonlinear term is written in the form

$$\theta \frac{\partial}{\partial x}(\tfrac{1}{2}u^2) + (1-\theta)u\frac{\partial u}{\partial x}, \qquad 0 \le \theta \le 1$$

and using central differences we obtain the finite element formulae above when

$\theta = 2/3$ Standard Galerkin

$\theta = 1$ Product Approximation

There are many possible finite element representations of the dispersion term, but in finite differences we always use

$$\frac{\partial^3 u}{\partial x^3} \sim \frac{1}{2h^3}[U_{m+2} - 2U_{m+1} + 2U_{m-1} - U_{m-2}]$$

3.4 Linearised Stability of Difference Schemes

We now carry out a standard Von Neumann stability analysis on the linearised forms of the various space and time discretisations of the KdV equation. Note that the linearisations do not involve the parameter θ. If E_m^n is the error in U_m^n where

$$E_m^n \sim e^{\gamma nk} e^{i\delta mh}, \qquad \gamma, \delta \text{ constants}$$

the Von Neumann condition is

$$|e^{\gamma k}| \le 1 . \tag{3.5}$$

If we put $A = \sin\phi(\alpha - 4\beta\sin^2\frac{\phi}{2})$ with $\phi = \delta h$, $\alpha = \frac{Uk}{h}$, $\beta = \frac{\varepsilon k}{h^3}$, where U is the constant solution about which the linearisation is performed we get the following results for the various time-stepping procedures.

Euler $|e^{\gamma k}| = \sqrt{1 + A^2}$ All modes unstable

Leap Frog $e^{\gamma k} = -iA \pm \sqrt{1 - A^2}$

(i) $A^2 \le 1$ Modes neutrally stable

(ii) $A^2 > 1$ Modes unstable

Crank-Nicolson $e^{\gamma k} = \frac{1 - 2iA}{1 + 2iA}$ All modes neutrally stable

3.5 Nonlinear Stability of Difference Schemes

Here the only results are for

$$\frac{\partial u}{\partial t} + \theta \frac{\partial}{\partial x}(\tfrac{1}{2}u^2) + (1-\theta)u\frac{\partial u}{\partial x} = 0 \qquad (\varepsilon = 0)$$

where for central differences in space, Fornberg [11] has shown that $\theta = 2/3$ is necessary for stability of leap frog and necessary and sufficient for stability of Crank Nicolson. Stability here means that the L_2 norm of the difference approximation does not increase faster in time than a fixed exponential function even if the mesh is refined.

3.6 Numerical Studies and Recurrence

The pure initial value problem for the KdV equation where solutions tend to zero rapidly as $|x|$ tends to ∞ causes no great problem numerically. In these problems after the elapse of some time, solitary waves (solitons) appear, each of which propagates with a fixed speed and unaltered shape. Analytically this problem can be solved by inverse scattering theory for potentials vanishing at $x = \pm\infty$ [12]. The initial value problem for solutions which are periodic in x is much more difficult theoretically and numerically, however, and this is the problem on which we carry out numerical tests. The initial numerical steps in this problem were taken by Zabusky and Kruskal [13] who observed that solutions with sinusoidal initial values developed into wave patterns which eventually reverted to a resemblance of the initial shape. These solutions periodic in space and almost periodic in time were more or less confirmed theoretically by Lax [14].

For our numerical experiments we choose the initial condition

$$u(x) = \cos 2\pi x , \qquad 0 \le x \le 1,$$

and periodic boundary conditions at $x = 0,1$ for $t > 0$. Since $\frac{\partial u}{\partial x} < 0$ for $0 < x < \frac{1}{2}$ at $t = 0$, instabilities were expected to appear initially and in order that these be not suppressed artificially, the explicit leap frog scheme with unstable and neutrally stable modes was used. A typical numerical experiment shows the sinusoidal initial condition developing a perturbation close to $x = \frac{1}{4}$ which grows into a wave pattern and then reverts to an approximation of the initial condition. This phenomenon is known as recurrence.

4. FOURIER ANALYSIS AND NON LINEAR INTERACTIONS

We consider a time dependent problem (e.g. KdV) with one space variable and periodic boundary conditions. The unit interval is divided equally by the points $x = mh$, $m = 0,1,\ldots,M$. The numerical solution at time t transforms into discrete Fourier space as

$$U_m(t) = \sum_{p=-M/2}^{+M/2} a_p(t) e^{2\pi i p \frac{m}{M}}, \tag{4.1}$$

where $|p| \le M/2$ is the wave number and M is even. From (4.1), we see that quadratic interactions lead to terms of the form

$$e^{2\pi i p_1 x}\, e^{2\pi i p_2 x} = e^{2\pi i (p_1 + p_2) x}$$

and if $p_1 + p_2 > M/2$, the newly formed mode is incorrectly represented as $p_1 + p_2 - M$, a phenomenon referred to as aliasing. That is, there is a transfer of energy from the mode with wavenumber $p_1 + p_2$ to that with wavenumber $p_1 + p_2 - M$.

4.1 Reduction of Dimension of Problem

The non linear interactions mentioned above can only be followed realistically when a small number of Fourier components is involved. For example, the problem with periodic boundary conditions and period M in space has the solution form

$$U(mh,t) = a(t)\sin\frac{2\pi m}{3} + b(t)\cos\frac{2\pi m}{3} + c(t) \qquad m = 0,1,2,\ldots \tag{4.2}$$

when $M = 3$, and

$$U(mh,t) = a(t)\sin\frac{\pi m}{2} + b_1(t)\cos\frac{\pi m}{2} + b_2(t)\cos \pi m + c(t) \qquad m = 0,1,2,3,\ldots \tag{4.3}$$

when $M = 4$.

The KdV equation is now written in the form

$$\frac{\partial u}{\partial t} + \theta \frac{\partial}{\partial x}(\tfrac{1}{2}u^2) + (1-\theta)u\frac{\partial u}{\partial x} + \varepsilon \frac{\partial^3 u}{\partial x^3} = 0, \qquad 0 \le \theta \le 1,\ \varepsilon > 0$$

and discretisation in space using central differences for the middle two terms leads to

$$\frac{\partial U}{\partial t} + \frac{1}{2h}(U_{m+1} - U_{m-1})[\tfrac{1}{2}\theta(U_{m+1} + U_{m-1}) + (1-\theta)U_m] + \frac{\varepsilon}{2h^3}[U_{m+2} - 2U_{m+1} + 2U_{m-1} - U_{m-2}] = 0 \tag{4.4}$$

Substitution of (4.2) into (4.4) leads to the 3-system

$$\frac{da}{dt} = \frac{\sqrt{3}}{4h}\left(1-\frac{3}{2}\theta\right)(a^2-b^2+2bc) - \varepsilon\,\frac{3\sqrt{3}}{2h^3}\,b$$
$$\frac{db}{dt} = -\frac{\sqrt{3}}{2h}\left(1-\frac{3}{2}\theta\right)a(b+c) + \varepsilon\,\frac{3\sqrt{3}}{2h^3}\,a \qquad (4.5)$$
$$\frac{dc}{dt} = 0$$

and substitution of (4.3) into (4.4) leads to the 4-system

$$\frac{da}{dt} = \frac{1}{h}\,b_1[2\theta b_2 - (b_2 - c)] - \frac{2\varepsilon}{h^3}\,b_1$$
$$\frac{db_1}{dt} = \frac{1}{h}\,a[2\theta b_2 - (b_2 + c)] + \frac{2\varepsilon}{h^3}\,a$$
$$\frac{db_2}{dt} = -\frac{1}{h}(1-\theta)ab_1 \qquad (4.6)$$
$$\frac{dc}{dt} = 0\,.$$

Systems similar to (4.5) and (4.6) were obtained by Briggs et al [15] for the case $\varepsilon = 0$. It should be noted that $\theta = \frac{2}{3}$ linearises the system (4.5) and $\theta = 1$ the system (4.6). Also (4.6) with $\varepsilon = 0$ and $\theta = 1$ appears on pp. 128, 129 of Richtmyer and Morton [16] as an example of a system which for appropriate initial conditions becomes unstable in time. If it were possible to solve the above semi-discrete systems exactly, no new modes would be introduced since aliasing projects modes formed by non linear interactions back on to one or other of the original modes. The amplitudes of the original modes then depend on the initial condition together with non linear interactions and aliasing.

Theoretical and numerical studies of systems (4.5) and (4.6) are being carried out at present.

4.2 Round-off Errors and Side Bands

The semi-discrete systems are now discretised in time by the leap frog scheme (mid point rule). It is unlikely that the nonlinear partial difference equations obtained will ever be solved exactly and so we resort to numerical methods. The introduction of round-off errors may of course lead to a focusing mechanism for the destabilisation of the nonlinear difference schemes, a phenomenon discussed in the important paper by Briggs et al [15]. We now relate this mechanism for the triggering of nonlinear instabilities to the disintegration of periodic wavetrains on deep water analysed in a classic paper by Benjamin and Feir [17]. To quote from the latter, consider a periodic wavetrain given by

$$u = ae^{i\zeta}, \qquad \zeta = kx - wt , \tag{4.7}$$

leading to

$$u \frac{\partial u}{\partial x} = a^2kie^{2i\zeta} ,$$

the second harmonic of the primary wave. Higher harmonics lead to amplitudes

$$a^3k^2, \; a^4k^3, \; \text{etc.},$$

which decrease only if a is <u>small</u>. Now consider <u>side band</u> modes to the primary mode of wave number k given by

$$\zeta_1 = (k+\mu)x - (w+\delta)t - \gamma_1 \qquad [\varepsilon_1]$$

and

$$\zeta_2 = (k-\mu)x - (w-\delta)t - \gamma_2 \qquad [\varepsilon_2]$$

where $[\varepsilon_1]$ and $[\varepsilon_2]$ are the respective small amplitudes. A simple calculation gives

$$\zeta_1 + \zeta_2 - 2\zeta = -(\gamma_1 + \gamma_2)$$

and so if

$$\gamma_1 + \gamma_2 = \text{constant},$$

called the <u>resonance condition</u>, the pair of side bands feed each other leading to <u>nonlinear instability</u>. Numerical experiments conducted by David Sloan [18] on $\frac{\partial u}{\partial t} + u \frac{\partial u}{\partial x} = 0$ show that the growth rate of a side band mode varies with the <u>separation</u>, in wave number space, between either side band and the primary mode.

In conclusion it should be stated that although there is a strong connection between round-off errors and side band modes, (4.7) is a solution of the KdV equation only after it has been linearised.

REFERENCES

1. MAY, R.M. Simple mathematical models with very complicated dynamics, Nature 261 (1976) 459-467.

2. FEIGENBAUM, M.J. The metric universal properties of period doubling bifurcations, Nonlinear dynamics (ed. R.H.G. Helleman), Annals N.Y. Acad. Sci. 357 (1980) 330-336.

3. HÉNON, M. A two-dimensional mapping with a strange attractor, Comm. Math. Phys. 50 (1976) 69-77.

4. SANZ-SERNA, J.M. Studies in numerical nonlinear instability I. Why do leapfrog schemes go unstable?, SIAM J. SCI. STAT. COMPUT. 6 (1985) (in the press).

5. HOPF, E. Bifurcation of a periodic solution from a stationary solution of a differential system, Ber. Math. Phys. Sachsische Acad. Wissensch. (Leipzig) 94 (1942) 1-22.

6. PRÜFER, M. Turbulence in Multistep methods for initial value problems, SIAM J. Appl. Math. 45 (1985) 32-69.

7. MITCHELL, A.R. and BRUCH, J.C. Jr. A numerical study of chaos in a reaction diffusion equation, Num. Meths. for P.D.Es. 1 (1985) 13-23.

8. WHITHAM, G.B. Linear and Nonlinear Waves, John Wiley and Sons, London, 1974.

9. LAX, P.D. and LEVERMORE, C.D. On the small dispersion limit for the KdV equation, Comm. Pure Appl. Math. 36, (1983) 253-290.

10. CHRISTIE, I., GRIFFITHS, D.F., MITCHELL, A.R. and SANZ-SERNA, J.M. Product approximation for nonlinear problems in the finite element method, IMA Journ. of Num. Anal. 1 (1981) 253-267.

11. FORNBERG, B. On the instability of leap-frog and Crank-Nicolson approximation of a non-linear partial differential equation, Math. Comp. 27 (1973) 45-57.

12. ZAKHAROV, V.E. and SHABAT, A.V. Exact theory of two-dimensional self-focusing and one dimensional self-modulation of waves in nonlinear media, Zh. Eksp. Tear. Fiz. 61 (1971) 118.

13. ZABUSKY, N.J. and KRUSKAL, M.D. Interaction of "solitons" in a collisionless plasma and the recurrence of initial states, Phys. Rev. Lett. 15 (1965) 240-243.

14. LAX, P.D. Periodic solutions of the KdV equation, Comm. Pure Appl. Maths. 28 (1975) 141-188.

15. BRIGGS, W., NEWELL, A.C. and SARIE, T. Focusing: A mechanism for instability of nonlinear finite difference equations, J. Comp. Phys. 51 (1983) 83-106.

16. RICHTMYER, R.D. and MORTON, K.W. Difference Methods for Initial Value Problems, John Wiley and Sons, New York, 1967.

17. BENJAMIN, T.B. and FEIR, J.E. The disintegration of wave trains on deep water, J. Fluid. Mech. 27 (1967) 417-430.

18. SLOAN, D.M. and MITCHELL, A.R. On nonlinear instabilities in leap-frog finite difference schemes, Num. Anal. Rep. NA/87, Univ. of Dundee, 1985.

A.R. Mitchell & D.F. Griffiths
Department of Mathematical Sciences
University of Dundee
Dundee DD1 4HN
Scotland
U.K.

K W MORTON & A PRIESTLY

On characteristic Galerkin and Lagrange Galerkin methods

1. INTRODUCTION

The exploitation of the characteristics in devising difference methods for hyperbolic equations has a long and distinguished history - see, for example, Courant, Isaacson & Rees [4] and Ansorge [1]. Similarly, Lagrangian co-ordinates have frequently been used to improve the treatment of convective terms in computational fluid dynamics. Generally, in both cases interpolation of a low order polynomial approximation is involved and leads to rather disappointing accuracy. More recently, however, these ideas have been combined with the Galerkin formulation to yield very powerful finite element procedures. It is the study of these methods, particularly as applied in two space dimensions, which is the subject of this paper.

To give a flavour of the methods and to put them in the context of familiar finite difference methods, consider the simple linear advection equation $\partial_t u + a\partial_x u = 0$, with constant positive characteristic speed a, on a uniform mesh of spacing Δx. Then piecewise constant elements, extending over $|x - j\Delta x| \leq \frac{1}{2}\Delta x$, and Euler time-stepping leads to the ECG (Euler characteristic Galerkin) method

$$U_i^{n+1} = (1 - \vartheta)U_{i-p}^n + \vartheta U_{i-p-1}^n \qquad (1.1a)$$

$$\text{where} \quad a\Delta t/\Delta x = \nu = p + \vartheta,\ 0 \leq \vartheta < 1. \qquad (1.1b)$$

This is the same first order accurate scheme as that yielded by piecewise linear interpolation in a characteristic difference method. Using continuous piecewise linear elements in the ECG method gives instead

$$(1 + \tfrac{1}{6}\delta^2)U_i^{n+1} = \left[(1 + \tfrac{1}{6}\delta^2) - \vartheta\Delta_o + \tfrac{1}{2}\vartheta^2\delta^2 - \tfrac{1}{6}\vartheta^3\delta^2\Delta_-\right]U_{i-p}^n, \qquad (1.2)$$

where we have used the usual difference operator notation

$$\Delta_o U_i = \tfrac{1}{2}(U_{i+1} - U_{i-1}),\ \delta^2 U_i = U_{i+1} - 2U_i + U_{i-1},\ \Delta_- = U_i - U_{i-1}.$$

Here the operator $(1 + \frac{1}{6}\delta^2)$ represents the familiar tridiagonal mass matrix obtained with linear elements: the scheme has third order accuracy. Intermediate between these are schemes based on piecewise constant elements but exploiting recovery techniques at each time step which use continuous piecewise linears - see Morton [8,9]: a typical example is

$$U_i^{n+1} = U_{i-p}^n - \tilde{\nu}\Delta_o \tilde{u}_{i-p}^n + \tfrac{1}{2}\tilde{\nu}^2 \delta^2 \tilde{u}_{i-p}^n \tag{1.3a}$$

where $$(1 + \tfrac{1}{8}\delta^2)\tilde{u}_i^n = U_i^n \tag{1.3b}$$

and $$\nu = p + \tilde{\nu}, \quad |\tilde{\nu}| \leq \tfrac{1}{2}. \tag{1.3c}$$

This is a second order accurate scheme.

The piecewise constant and piecewise linear basis functions used here correspond to the zeroth order and first order B-splines generated from the characteristic function $\chi^{(1)}(t)$ of the interval $[-\frac{1}{2},\frac{1}{2}]$ by the convolution recursion

$$\chi^{(m+1)}(t) = (\chi^{(m)} * \chi^{(1)})(t) = \int \chi^{(m)}(s)\chi^{(1)}(s-t)ds. \tag{1.4}$$

The ECG schemes are generated by translating the approximation U^n at the old time level through ν mesh intervals before projecting on to the approximation space to obtain the new approximation U^{n+1}. This also involves a convolution and some of the key properties of the schemes stem from this correspondence. Thus a scheme based on $\chi^{(m)}(t)$ without recovery can be written

$$\sum_{(k)} \chi^{(2m)}(k) U_{i+k}^{n+1} = \sum_{(k)} \chi^{(2m)}(k+\nu) U_{i+k}^n, \tag{1.5}$$

where the coefficients $\chi^{(2m)}(k)$ on the left are generated by the convolutions involved in calculating the mass matrix and the $\chi^{(2m)}(k+\nu)$ on the right from the translation followed by projection. On the other hand if recovery using basis functions $\chi^{(m+p)}(t)$ is used, the two stage process can be written

$$\sum_{(k)} \chi^{(2m+p)}(k) \tilde{u}_{i+k}^n = \sum_{(k)} \chi^{(2m)}(k) U_{i+k}^n \tag{1.6a}$$

$$\sum_{(k)} \chi^{(2m)}(k) U_{i+k}^{n+1} = \sum_{(k)} \chi^{(2m+p)}(k+\nu) \tilde{u}_{i+k}^n. \tag{1.6b}$$

Thus for p = 2 the recovered scheme is equivalent to an unrecovered scheme using basis functions $\chi^{(m+1)}(t)$: in particular the scheme with piecewise constants if combined with recovery using quadratic splines yields the scheme (1.2) based on piecewise linears.

Stability and accuracy is best analysed by substituting a Fourier mode $U_j^n = \lambda^n e^{ij\xi}$. Then one finds for (1.5)

$$\lambda(\xi,\nu) = \sum_{(k)}\chi^{(2m)}(k+\nu)e^{ik\xi}/\sum_{(k)}\chi^{(2m)}(k)e^{ik\xi}. \tag{1.7}$$

For m = 2 which corresponds to (1.2), the difference from the exact result is given by

$$e^{-i\nu\xi} - \lambda(\xi,\nu) \sim \frac{1}{24}\nu^2(1-\nu)^2\,\xi^4\,e^{-ip\xi} \text{ as } \xi \to 0: \tag{1.8}$$

and for stability one finds

$$|\lambda|^2 = 1 - (4/3)r^2[1+(4/3)r]/q^2, \tag{1.9}$$

where $r = \nu(1-\nu)\sin^2\frac{1}{2}\xi$, $q = 1-(2/3)\sin^2\frac{1}{2}\xi$, which implies that $|\lambda| \leq 1$ for $\nu \in [-\frac{1}{2}, \frac{3}{2}]$. Thus with ν limited to the range [0,1) and the scheme exact for integral values of ν, one sees that there is a valuable stability margin implied by (1.9) and that the maximum value of the local error (1.8) is only $(1/384)\xi^4$ at $\nu = \frac{1}{2}$. It is worth noting that (1.2) is a member of a one-parameter family of schemes obtained by projecting with the mixed norm

$$\langle u,v\rangle + \gamma^2\langle\partial_x u,\partial_x v\rangle, \tag{1.10}$$

where $\langle\cdot,\cdot\rangle$ denotes the L^2 inner product, and including several well-known third order difference schemes: the local error for these satisfy

$$e^{-i\nu\xi} - \lambda^{(\gamma)}(\xi,\nu) \sim \frac{1}{24}\nu(1-\nu)[\nu(1-\nu) + 12(\gamma/\Delta x)^2]e^{-ip\xi} \tag{1.11}$$

which demonstrates that the L^2 projection gives the best result in this case - see Morton [9].

2. ALTERNATIVE FORMULATIONS

Let us consider first the Cauchy problem for the linear advection equation for $u(\underline{x},t)$

$$\partial_t u + \underline{a}.\nabla u = 0, \quad u(0) = u_0, \tag{2.1}$$

where the velocity field $\underline{a}(\underline{x})$ is incompressible,

$$\underline{\nabla}\cdot\underline{a} = 0. \tag{2.2}$$

We define the translation operator $T(\tau)$ along the trajectories generated by $\underline{a}(\underline{x})$:-

$$T(\tau)\underline{x} = \underline{X}(\tau), \text{ where } \underline{X}(0) = \underline{x}, \; d\underline{X}/dt = \underline{a}(\underline{X}); \tag{2.3}$$

hence the relation

$$u(T(\tau)\underline{x}, t+\tau) = u(\underline{x}, t) \tag{2.4}$$

gives the solution to (2.1).

Characteristic Galerkin and Lagrange Galerkin methods are indistinguishable in this case. The most direct formulation for an approximation given in terms of basis functions $\phi_j(\underline{x})$,

$$U^n(\underline{x}) = \sum U_j^n \phi_j(\underline{x}) \tag{2.5}$$

uses (2.4) directly to obtain

$$\langle U^{n+1}, \phi_i \rangle = \int U^n(\underline{x})\phi_i(\underline{y})d\underline{y} \;, \; \underline{y} = T(\Delta t)\underline{x}. \tag{2.6}$$

We call this the direct approach and it is the same as that used by several authors - Bercovier et al. [3], Douglas & Russell [5], for example. An alternative, weak formulation was introduced by Benqué et al. [2]. It introduces test functions $\psi_i(\underline{x}, t)$ and from

$$\int_t^{t+\Delta t} \langle \partial_t u + \underline{a}.\underline{\nabla} u, \psi_i \rangle dt = 0 \tag{2.7a}$$

deduces the identity

$$\langle u(\cdot, t+\Delta t), \psi_i(\cdot, t+\Delta t)\rangle - \langle u(\cdot, t), \psi_i(\cdot, t)\rangle$$

$$= \int_t^{t+\Delta t} \langle u, \partial_t \psi_i + \underline{\nabla}.(\underline{a}\psi_i)\rangle dt. \tag{2.7b}$$

Now, by the incompressibility condition, $\underline{\nabla}.(\underline{a}\psi_i) = \underline{a}.\underline{\nabla}\psi_i$; so this last term can be set to zero if the test functions are defined by

$$\psi_i(T(\tau)\underline{x}, t+\tau) = \psi_i(\underline{x}, t). \tag{2.7c}$$

If in addition we set

$$\psi_i(\underline{y}, t+\Delta t) = \phi_i(\underline{y}) \tag{2.7d}$$

and substitute U into (2.7b) we obtain

$$\langle U^{n+1}, \phi_i \rangle = \int U^n(\underline{x})\phi_i(\underline{y})d\underline{x}\ , \ \underline{y} = T(\Delta t)\underline{x}. \tag{2.8}$$

Comparing with (2.6) we see that with exact integration we have exactly the same scheme since

$$\underline{\nabla}\cdot\underline{a} = 0 \Rightarrow d\underline{y} = d\underline{x}, \text{ where } \underline{y} = T(\tau)\underline{x}. \tag{2.9}$$

The third formulation we shall consider is that used by Morton [7,8,9]. This, like the direct method, uses (2.4) to characterise the solution, but then manipulates (2.6) to recover a weak form of the differential equation: for the exact solution at the two time-steps we obtain

$$\begin{aligned}\langle u^{n+1}, \phi_i \rangle - \langle u^n, \phi_i \rangle &= \int u^{n+1}(\underline{y})\phi_i(\underline{y})d\underline{y} - \int u^n(\underline{x})\phi_i(\underline{x})d\underline{x} \\ &= \int u^n(\underline{x})[\phi_i(\underline{y})d\underline{y} - \phi_i(\underline{x})d\underline{x}]\end{aligned} \tag{2.10}$$

where again $\underline{y} = T(\Delta t)\underline{x}$; now introduce the test function which is the average of ϕ_i along the trajectory (s is distance along trajectory)

$$\Phi_i(\underline{x}) = \frac{1}{s(\underline{y},\underline{x})}\int_{\underline{x}}^{\underline{y}} \phi_i(T(\tau)\underline{x})ds; \tag{2.11a}$$

then

$$\Delta t\underline{a}\cdot\underline{\nabla}\Phi_i d\underline{x} = \phi_i(\underline{y})d\underline{y} - \phi_i(\underline{x})d\underline{x} \tag{2.11b}$$

so that (2.10) becomes after integration by parts

$$\langle u^{n+1} - u^n, \phi_i\rangle + \Delta t\langle \underline{a}.\underline{\nabla} u^n, \phi_i\rangle = 0. \tag{2.12}$$

This identity yields the ECG scheme

$$\langle U^{n+1} - U^n, \phi_i\rangle + \Delta t\langle \underline{a}.\underline{\nabla} U^n, \phi_i\rangle = 0, \tag{2.13}$$

which with exact integration is identical with (2.6) and (2.8). We will always assume that $\sum \phi_i \equiv 1$, so that exact conservation of $\int U dx$ follows immediately from any of the formulations. Suppose also that we denote by S_Δ the solution operator $u(t+\Delta t) = S_\Delta u(t)$ for the equation (2.1) over the time-step Δt, that is from (2.3) and (2.4),

$$(S_\Delta v)(\underline{y}) = v(T(-\Delta t)\underline{y}). \tag{2.14}$$

Then because of (2.9), for the L^2 norm we have

$$\|S_\Delta\| = 1 = \|S_\Delta^{-1}\|. \tag{2.15}$$

Thus the unconditional stability of the direct or weak formulations follow immediately:-

$$\begin{aligned} &\langle U^{n+1}, \phi_i\rangle = \langle S_\Delta U^n, \phi_i\rangle \\ \Rightarrow\ &\|U^{n+1}\|^2 = \langle S_\Delta U^n, U^{n+1}\rangle \le \|S_\Delta U^n\|\,\|U^{n+1}\| \\ \Rightarrow\ &\|U^{n+1}\| \le \|U^n\|; \end{aligned} \tag{2.16a}$$

similarly

$$\begin{aligned} &\langle U^{n+1}, \phi_i\rangle = \langle U^n, S_\Delta^{-1}\phi_i\rangle \\ \Rightarrow\ &\|U^{n+1}\|^2 = \langle U^n, S_\Delta^{-1}U^{n+1}\rangle \le \|U^n\|\,\|S_\Delta^{-1}U^{n+1}\| \\ \Rightarrow\ &\|U^{n+1}\| \le \|U^n\|. \end{aligned} \tag{2.16b}$$

For the ECG scheme (2.13) we write, after integrating the second term by parts,

$$\|U^{n+1}\|^2 = \langle U^n, U^{n+1} + \Delta t\underline{a}.\underline{\nabla}(\sum U_i^{n+1}\phi_i)\rangle \tag{2.17}$$

and exploit the fact that

$$\|V + \Delta t\underline{a}.\underline{\nabla}(\sum V_i\phi_i)\| \leq \|V\| \tag{2.18}$$

to deduce unconditional stability. The slightly different forms of these arguments will of course be important when the schemes have to be approximated through the use of quadrature.

The schemes will also differ for more general problems than (2.1). The first two have been developed for relatively smooth problems, such as diffusion convection and the Navier-Stokes equations, while the latter has been developed through shock-modelling problems. Thus it is difficult to choose one problem to demonstrate the key features of all three. Consider first then

$$\partial_t u + \underline{a}.\underline{\nabla}u = h(\underline{x},t,u), \tag{2.19}$$

where $\underline{a}$ is not necessarily incompressible. The direct Lagrange Galerkin method typically leads to

$$\langle U^{n+1} - S_\Delta U^n, \phi_i\rangle = \int_0^{\Delta t}\int h(T((-\tau)\underline{y}, t_{n+1} - \tau,\ S_\Delta U^n(\underline{y}))\phi_j(\underline{y})d\underline{y}d\tau, \tag{2.20}$$

where S_Δ is defined as before by (2.3) and (2.14). The weak formulation on the other hand gives

$$\langle U^{n+1},\phi_i\rangle - \langle U^n, S_\Delta^{-1}\phi_i\rangle = \int_0^{\Delta t}\int[U^n\underline{\nabla}.\underline{a} + h(T(\tau)\underline{x},\ t_n+\tau,\ U^n(\underline{x}))]\phi_j(\underline{y})d\underline{x}d\tau: \tag{2.21}$$

the extra term here comes from the fact that the mapping $T(\tau)$ is no longer area-preserving and indeed, if J is the Jacobian of the transformation,

$$\frac{d}{dt}\det J = \underline{\nabla}.\underline{a}\ \det J. \tag{2.22}$$

The main difference is that in the first case the emphasis is on integration over the elements at the new time level and in the second on integration at the old time level.

In (2.19)-(2.21), the velocity field $\underline{a}$ may of course depend on u: then it will normally be $\underline{a}(U^n)$ that is used to define S_Δ and $T(\tau)$. However, especially for systems of equations, one needs to distinguish between what is often a fluid velocity field and a characteristic velocity. Thus for the system of conservation laws in one dimension

$$\partial_t \underset{\sim}{w} + \partial_x \underset{\sim}{f}(\underset{\sim}{w}) = \underset{\sim}{h}(x,t,\underset{\sim}{w}) \tag{2.23}$$

the Jacobian $A(\underset{\sim}{w}) = \partial\underset{\sim}{f}/\partial\underset{\sim}{w}$ has eigenvalues which often take the form u + c where the velocity u is common to all of them. Then we refer to Lagrange Galerkin methods when only u is used to define S_Δ and $T(\tau)$, with the other terms being treated as in (2.20), (2.21) by the purely Galerkin scheme. In the characteristic Galerkin methods, on the other hand, the full characteristic speed is used: thus for the scalar version of (2.23) the ECG scheme gives

$$<U^{n+1} - U^n, \phi_i> + \Delta t<\partial_x f(U^n), \Phi_i^n> = \Delta t<h(\cdot, t_n, U^n), \Phi_i^n> \tag{2.24}$$

where Φ_i^n refers to (2.11a) defined using $\partial f/\partial u$ evaluated at U^n: for a system of equations see Morton & Sweby [10].

In all three schemes, a diffusion term $\partial_x(D\partial_x u)$ will be approximated by the usual Galerkin terms $<D\partial_x U, \partial_x \phi_i>$, with an explicit or implicit choice of U depending on the severity of the consequential stability limit.

3. CHOICE OF QUADRATURE SCHEMES

Some form of quadrature is generally needed both for the inner products on the right of (2.6), (2.8) or (2.13) and either for the trajectories needed in the first two or the special test function Φ_i in the last. A bad choice can destroy the key properties of the schemes, not only the accuracy, but also the stability. We begin by considering advection by an incompressible velocity field, as described by (2.1), (2.2), and either linear elements on triangles or bilinear elements on rectangles.

In all three methods the mass matrix is needed and a linear system involving it has to be solved: there is no difficulty in computing this exactly, although there may sometimes be an advantage to using the same quadrature scheme on it, and we assume that the linear system is solved with adequate accuracy. We have found a preconditioned conjugate gradient method to be very efficient, requiring only 3 iterations and normally taking less than 10% of the total time for linear advection.

Thus for the direct and weak methods there remains the problem of calculating the right-hand side vector which we will denote by $\{R_i\}$. Abbreviating an approximation to $T(\Delta t)$ defined in (2.3) by T_h and to $T(-\Delta t)$

by T_h^{-1}, any standard quadrature applied to the direct method (2.6) will yield an approximation

$$R_i \approx \sum_{(e)}\sum_{(k)} w^{(k)} \phi_i(\underline{x}^{(k)}) U^n(T_h^{-1}\, \underline{x}^{(k)}): \tag{3.1}$$

here the sum $\sum_{(e)}$ is over all the elements, $\underline{x}^{(k)}$ are the quadrature points in the elements and $w^{(k)}$ the corresponding quadrature weights. On the other hand the weak method (2.8) gives

$$R_i \approx \sum_{(e)}\sum_{(k)} w^{(k)} \phi_i(T_h\underline{x}^{(k)}) U^n(\underline{x}^{(k)}). \tag{3.2}$$

Regarding the conservation of $\int U d\underline{x}$, some advantage would seem to lie with the weak method since summing (3.2) over i immediately gives exact conservation, assuming that the quadrature rule is exact for linear or bilinear functions: on the other hand this does not follow as readily from (3.1). For stability, however, there seems little to choose between them: assuming that the quadrature rule is exact for the mass matrix or it is used anyway, and denoting by $\langle\cdot,\cdot\rangle_h$ the resulting inner product, stability will follow if for every U, V in the trial space there holds

$$\text{(direct)} \quad \langle S_h U, V\rangle_h \leq [1 + O(\Delta t)] \|U\|_h \|V\|_h$$

$$\text{or (weak)} \quad \langle U, S_h^{-1} V\rangle_h \leq [1 + O(\Delta t)] \|U\|_h \|V\|_h, \tag{3.3b}$$

where S_h, S_h^{-1} correspond to (2.14) using T_h, T_h^{-1}.

Even the one dimensional case with constant advection velocity on a uniform mesh quickly shows that too simple a quadrature rule is inadequate. For example, using the centroid only in (3.1) or (3.2) and the exact mass matrix gives stability only if $\vartheta \in [0, 1/\sqrt{6}] \cup [1 - 1/\sqrt{6}, 1]$: with the mass matrix also evaluated by centroid quadrature it is unconditionally unstable - because the mass matrix is then not strictly positive. On the other hand using the nodes for quadrature and the exact mass matrix is unconditionally unstable: while using nodal quadrature for the mass matrix here causes the scheme to degenerate to the 1st order scheme (1.1) which is unconditionally stable but inaccurate. Combining these to give Simpson's rule still leaves a gap in the stability range. Note that we have not distinguished between the direct and weak method here because the following simple calculation shows

that they are identical if both the canonical basis function $\phi(s)$ and the quadrature rule are symmetric:-
the coefficient of U_j^n in the direct method is proportional to

$$\sum_{(e)}\sum_{(k)} w^{(k)}\phi(s^{(k)} + e - i)\phi(s^{(k)} + e - j - \nu); \tag{3.4a}$$

the corresponding sum for the weak method is

$$\sum_{(e')}\sum_{(k')} w^{(k')}\phi(s^{(k')} + e' - i + \nu)\phi(s^{(k')} + e' - j)$$

$$= \sum_{(e')}\sum_{(k')} w^{(k')}\phi(-s^{(k')} - e' + j)\phi(-s^{(k')} - e' + i - \nu) \tag{3.4b}$$

which is the same as (3.4a) if we put $s_k' = 1 - s_k$, $e' = i + j - e - 1$. Unfortunately, this seems to be one of very few simple results regarding the effect of quadrature: if the centroid scheme is continued as a sequence of Gaussian quadrature schemes with 1, 2, 4, 8... points, certainly for 8 points the results are indistinguishable from the exact results; but the problem of possible instability regions is unclear, with those for the 2 point scheme lying near the ends of the interval [0, 1] in contrast to the centroid case where we have seen it is unstable in the centre.

In the two dimensional case we have largely to rely on the results of numerical experiment: for the bilinear case on rectangles, with a constant advection velocity, Fourier analysis can be applied to the tensor product basis functions so that the truncation error for exact integration given in (1.8) is replaced by a sum dependent on the two components ν_x, ν_y. This could also be done for product quadrature formulae, but for linear elements on triangles no such results are available. We have therefore used a standard test in which an initial cone centred on $(-\frac{1}{2}, 0)$ and of radius $\frac{1}{4}$ is advected in a circle around the origin with velocity vector $\underline{a} = 2\pi(-y, x)$: we have found it most illuminating to use $\sin^2 4\pi r$ as the cone form and have typically taken $\Delta x = \Delta y = 1/16$. Then typical results are summarised in the table below: these are for 25 steps of $\Delta t = 0.02$, which takes one half way round the circle, and for piecewise linears on triangles: the initial data was interpolated rather than L^2 fitted. One sees most of the earlier deductions borne out: the weak formulation conserves $\int U d\underline{x}$ more accurately, though the solution is generally less smooth; the mid-edge quadrature shows some tendency to instability like the nodal quadrature (which is quite unstable); seven Gauss points are needed for reliable results, with the weak and direct formulations agreeing. Other tests confirm the instability of the lower order quadrature schemes for various choices of time step.

TABLE Errors with rotating cone problems: $\nu = 0.32\pi$ at centre.

Quadrature	Formulation	$\int U d\underline{x}$	Max	Min	ℓ_2 error
Centroid	Weak	14.965	0.611	-0.014	1.2
Centroid	Direct	14.938	0.608	-0.007	1.1
Mid edge	Weak	14.967	0.930	-0.047	1.6×10^{-1}
Mid edge	Direct	14.919	0.933	-0.043	7.9×10^{-2}
3 Gauss	Weak	14.961	1.034	-0.025	4.8×10^{-2}
3 Gauss	Direct	15.006	1.010	-0.024	1.5×10^{-2}
7 Gauss	Weak	14.964	0.996	-0.014	2.2×10^{-2}
7 Gauss	Direct	14.965	0.991	-0.013	2.0×10^{-2}

Quadrature schemes used by Benqué et al [2] are hybrids between the direct and weak formulations. The nodes of triangles at the $U^{n+1}(\underline{y})$ mesh are mapped back to the $U^n(\underline{x})$ mesh: the integration is then carried out over these transformed triangles by using isoparametric transformations to canonical triangles and standard quadrature rules. Thus the integration elements are $d\underline{x}$, as in the weak formulation, while the quadrature over gradient discontinuities is for U^n, as in the direct formulation: our experiments show that, though motivation came from the former, the results are indistinguishable from those of the latter.

Finally then we come to quadrature rules for the ECG formulation. Some schemes for these were given by Morton [7] and the conclusions of our more extensive tests are consistent with these. The most important is that, although the test function $\Phi(\underline{x})$ may be approximated fairly crudely, it is important to evaluate the inner product involving it very accurately. Thus in [7] various schemes were proposed in which $\Phi_i^n(\underline{x})$ was approximated by linear combinations of $\phi_j(\underline{x})$, with the nodes j being near neighbours of the nodes i, and also of $\underline{\nabla}\phi_j(\underline{x})$: then all of the resulting inner products are on one mesh and can be accurately evaluated by any reasonable quadrature method. We have also used schemes in which the integral (2.11a) is evaluated by subdividing the $(\underline{x}, \underline{y})$ paths and using the trapezoidal rule on each subinterval: even with no subdivision, good results could be obtained but then the inner product $<\partial_x f, \phi_i^n>$ needed to be evaluated by a very accurate method as with the direct and weak formulations; with the centroid rule for example the stability restriction was approximately $\nu \lesssim \frac{1}{4}$, although the accuracy was then very high.

To conclude this section, great care needs to be taken with the quadrature when linear or bilinear elements are used. However, the accuracy that can then be achieved is very high indeed, as can be seen from the last two rows of the Table, and the results published by other authors [2,3].

4. ECG SCHEMES USING RECOVERY TECHNIQUES

The difficulties with using piecewise linear elements on a fixed mesh are not confined to those of quadrature. For hyperbolic problems with severe non-linearities they typically give oscillatory approximations which can generate non-linear instability. The situation is very similar to that with difference methods: in this case, for shock modelling there has been a move away from conventional second order schemes such as Lax-Wendroff towards upwind methods which are based on first order schemes, with second order terms introduced only in so far as they do not generate oscillations. In the same spirit Morton and Sweby [10] have developed the ECG scheme based on piecewise constant elements for one-dimensional shock modelling problems: improved accuracy (second order in smooth parts of the flow) is obtained by using an adaptive recovery technique, a simple example of which was given in (1.3).

Consider then the scalar conservation law

$$\partial_t u + \partial_x f(u) = 0. \tag{4.1}$$

Suppose $U^n(x)$ is piecewise constant and that at each time step an approximation $\tilde{u}^n(x)$ is recovered which more accurately reflects the important features of the solution, but which has U^n as its projection:-

$$<U^n - \tilde{u}^n, \phi_i> = 0 \qquad \forall\ \phi_i. \tag{4.2}$$

Then the ECG scheme based on this recovery technique is given by, cf. (2.24),

$$<U^{n+1} - U^n, \phi_i> + \Delta t <\partial_x f(\tilde{u}^n), \tilde{\Phi}_i^n> = 0, \tag{4.3a}$$

where

$$\tilde{\Phi}_i^n(x) = \frac{1}{a(\tilde{u}^n(x))\Delta t} \int_x^{x+a(\tilde{u}^n)\Delta t} \phi_i(z)dz. \tag{4.3b}$$

If $\tilde{u}^n$ is constructed to be piecewise linear (but possibly discontinuous) the great advantage is that all the integrals in (4.3) can be explicitly calculated to give a very straightforward algorithm.

If there is no recovery, if the flux function has the single sonic point $\bar{u}$ at which $a(\bar{u}) = (\partial f/\partial u)|_{\bar{u}} = 0$ and if $|\nu| = |a\Delta t/\Delta x| \leq 1$, the ECG scheme reduces to the well-known Engquist-Osher algorithm [6]: in the form that can be derived most obviously from (4.3), U^{n+1} is obtained from U^n by distributing the increments $[f(U^n_{k+1}) - f(U^n_k)]$ as follows:-

$$\underline{\text{add}}\ \frac{-\Delta t}{\Delta x}[f(U^n_{k+1})-f(\bar{u})]\ \underline{\text{to}}\ \begin{cases} U^n_{k+1} \\ U^n_k \end{cases} \underline{\text{if}}\ \begin{cases} a(U^n_{k+1}) > 0 \\ a(U^n_{k+1}) < 0 \end{cases} \tag{4.4a}$$

and

$$\underline{\text{add}}\ \frac{-\Delta t}{\Delta x}[f(\bar{u})-f(U^n_k)]\ \underline{\text{to}}\ \begin{cases} U^n_{k+1} \\ U^n_k \end{cases} \underline{\text{if}}\ \begin{cases} a(U^n_k) > 0 \\ a(U^n_k) < 0. \end{cases} \tag{4.4b}$$

This satisfies an entropy condition and can thus be shown to converge to the correct physical solution.

To improve the accuracy the approximation $\tilde{u}^n$ is recovered from U^n by using

(i) the projection property (4.2),

(ii) neighbouring values of U^n_j if justified by smoothness criteria,

(iii) known properties of the exact solution, such as monotonicity, positivity etc..

The main smoothness discriminant used in [8,9,10] has been the difference ratio

$$r_k = (U^n_k - U^n_{k-1})/(U^n_{k+1} - U^n_k); \tag{4.5}$$

and the main feature of solutions to (4.1) that has been exploited is that they are piecewise smooth (without oscillations) and contain jumps (shocks and contact discontinuities). Thus the recovery procedure has consisted of two parts:-

(i) each cell is scanned and a jump is "recognised" in element k if

$$a(U^n_{k-1}) \geq a(U^n_{k+1}) \text{ and}$$
$$r_k > 0,\ |r_{k-1}| \ll 1,\ |r_{k+1}| \gg 1\ ; \tag{4.6a}$$

(ii) if there is no jump recognised in either cell k or k+1, the discontinuity at $x_{k+\frac{1}{2}}$ is resolved by a piecewise linear segment extending $\frac{1}{2}\theta_{k+\frac{1}{2}}\Delta x$ either side, between values $\tilde{u}_k^n$ and $\tilde{u}_{k+1}^n$: the set of $\{\theta_{k+\frac{1}{2}}\}$ is chosen so that

$$\Delta_+\tilde{u}_k^n \text{ has the same sign as } \Delta_+U_k^n. \tag{4.7}$$

Condition (4.7) ensures that no new extrema are introduced, although existing extrema in U^n may be enhanced - indeed, this is the key to improved accuracy in non-monotone solutions. Detailed algorithms for selecting $\{\theta_{k+\frac{1}{2}}\}$ are given in [10]; then the following tridiagonal system for the $\{\tilde{u}_j^n\}$

$$\frac{1}{8}\theta_{k-\frac{1}{2}}\tilde{u}_{k-1}^n + (1-\frac{1}{8}\theta_{k-\frac{1}{2}}-\frac{1}{8}\theta_{k+\frac{1}{2}})\tilde{u}_k^n + \frac{1}{8}\theta_{k+\frac{1}{2}}\tilde{u}_{k+1}^n = U_k^n \tag{4.8}$$

results from (4.2).

The update procedure resulting from (4.3) with this form of $\tilde{u}^n$ turns out to be very simple. The relative position of each jump in its element follows from (4.2) as

$$(1-\eta)\tilde{u}_{k+1}^n + \eta\tilde{u}_k^n = U_k^n \tag{4.9a}$$

and this is then moved with the jump speed

$$[f(\tilde{u}_{k+1}^n)-f(\tilde{u}_{k-1}^n)]/[\tilde{u}_{k+1}^n-\tilde{u}_{k-1}^n]. \tag{4.9b}$$

Moreover, the linear sections, centred at $\tilde{u}_{k+\frac{1}{2}}$ with slope $m_{k+\frac{1}{2}}$, can each be treated by an Engquist-Osher algorithm of the form (4.4) but with a locally modified flux function. Thus we define

$$F_{k+\frac{1}{2}}(\tilde{u}) = f(\tilde{u}) - f(\tilde{u}_{k+\frac{1}{2}}) + \tfrac{1}{2}(\tilde{u}-\tilde{u}_{k+\frac{1}{2}})^2/m_{k+\frac{1}{2}}\Delta t \tag{4.10}$$

together with $A_{k+\frac{1}{2}}(\tilde{u}) = \partial F_{k+\frac{1}{2}}/\partial\tilde{u}$: and we denote by $u_{k+\frac{1}{2}}^{(\ell)}$, $\ell=1,2,\ldots,m$ the sonic points of $A_{k+\frac{1}{2}}$ between $\tilde{u}_k^n$ and $\tilde{u}_{k+1}^n$, which it is convenient to call $u_{k+\frac{1}{2}}^{(0)}$, $u_{k+\frac{1}{2}}^{(m+1)}$ respectively. Then for $|\nu| \leq \frac{1}{2}$, the update procedure for $\{U_j^{n+1}\}$ is based on adding to $\{\tilde{u}_j^n\}$ the following increments for each value of k:-

for $\ell=0,1,\ldots,m$

$$\underline{\text{add}}\ -\frac{\Delta t}{\Delta x}[F_{k+\frac{1}{2}}(u_{k+\frac{1}{2}}^{(\ell+1)})-F_{k+\frac{1}{2}}(u_{k+\frac{1}{2}}^{(\ell)})]\ \underline{\text{to}}\ \begin{cases}\tilde{u}_{k+1}^n \\ \tilde{u}_k^n\end{cases}\ \underline{\text{if}}\ \begin{cases}A_{k+\frac{1}{2}}(\tilde{u}) > 0 \\ A_{k+\frac{1}{2}}(\tilde{u}) < 0\end{cases} \tag{4.11}$$

the sign of $A_{k+\frac{1}{2}}$ holding between the pair of sonic points involved.

This algorithm has been extended to systems of equations in [10] where the results of various numerical tests are also given: these show that the method is very comparable in accuracy with the best flux-limited difference schemes - sometimes rather better, sometimes not quite so good. Taken together with the results in [2,3,5] and other publications of these authors, they demonstrate the very wide range of success achieved by the characteristic Galerkin and Lagrange Galerkin approach: from complicated, multidimensional (but low flow speed) industrial flow problems to shock-modelling in one dimension. One of our aims in this work, where we have brought together the methodologies that have been developed for the two types of problem, has been to prepare the way for tackling unsteady multidimensional shock-modelling.

REFERENCES

1. ANSORGE, R. Numer. Math. 5 (1963) 443.

2. BENQUÉ, J.P., LABADIE, G. & RONAT J. A new finite element method for Navier-Stokes equations coupled with a temperature equation. Proc. 4th Int. Symp. on Finite Element Meth. in Flow Problems, Chuo University, Tokyo, 1982 (Ed. Tadahiko Kawai).

3. BERCOVIER, M., PIRONNEAU, O., HASBANI, Y. & LIVNE, E. Characteristic and finite element methods applied to the equations of fluids, Procs. MAFELAP 1981 Conf. (Ed. WHITEMAN, J.R.) 471-478, Academic Press, London 1982.

4. COURANT, R., ISAACSON, E. & REES, M. On the solution of non-linear hyperbolic differential equations by finite differences, Comm. Pure & Appl. Math. 5 (1952) 243-264.

5. DOUGLAS Jr. J. & RUSSELL, T.F. Numerical methods for convection-dominated problems based on combining the method of characteristics with finite element or finite difference procedures, SIAM J. Numer. Anal. 19 (1982) 871-885.

6. ENGQUIST, B. & OSHER, S. One sided difference equations for non-linear conservation laws, Math. Comp. 36 (1981) 321-352.

7. MORTON, K.W. Generalised Galerkin methods for steady and unsteady Problems, Proc. IMA Conf. on Numer. Meth. for Fluid Dynamics (Eds. MORTON, K.W. & BAINES, M.J.) 1-32, Academic Press, 1982.

8. MORTON, K.W. Characteristic Galerkin methods for hyperbolic problems, Proc. 5th GAMM Conf. on Numer. Meth. in Fluid Mech. (Eds. PANDOLFI, M. & PIVA, R.) 243-250, F. Vieweg & Sohn, 1983.

9. MORTON, K.W. Generalised Galerkin methods for hyperbolic problems, Oxford Univ. Comp. Lab. Report 84/1 (1984), (to appear in Appl. Mech. Engng.).

10. MORTON, K.W. & SWEBY, P.K. A comparison of flux-limited difference methods and characteristic Galerkin methods for shock modelling, Oxford Univ. Comp. Lab. Report 85/3 (1985).

K.W. Morton and A. Priestley
Oxford University Computing Laboratory
Numerical Analysis Group
8-11 Keble Road
Oxford OX1 3QD

The work reported here forms part of the research programme of the Oxford/Reading Institute for Computational Fluid Dynamics.

A H G RINNOOY KAN & G T TIMMER

The multi-level single linkage method for unconstrained and constrained global optimization

1. INTRODUCTION

The global optimization problem is to find the global optimum (say, the global minimum) x_* of a real valued twice continuously differentiable objective function $f: \mathbb{R}^n \to \mathbb{R}$. For computational reasons one usually assumes that a set $S \subset \mathbb{R}^n$ which is convex, compact and contains the global minimum as an interior point, is specified in advance. Nevertheless, the problem to find

$$y_* = \min_{x \in S}\{f(x)\} \tag{1.1}$$

remains essentially one of unconstrained optimization.

Only a few solution methods for this problem have been developed so far, certainly in comparison with the multitude of nonlinear programming methods that aim for an arbitrary local minimum. For a survey of global optimization methods we refer to Dixon and Szegö [10], Dixon and Szego [11], Rinnooy Kan and Timmer [23] and Timmer [27]. It appears not to be possible to design methods which offer an absolute guarantee of success for arbitrary f. Therefore, most methods are of a stochastic nature and provide an asymptotic guarantee in a stochastic sense. For instance, if the function is evaluated in points which are drawn from a uniform distribution over S, then it can be shown that the smallest function value found converges to the global minimum value y_* with probability 1 (i.e. almost surely) (cf. Rubenstein [26]).

In Section 2, we describe a folklore method in this category, known as Multistart, paying particular attention to an appropriate stopping rule for this method. The Multi Level Single Linkage method eliminates the inherent inefficiencies of Multistart while retaining its theoretical properties; it is briefly described in Section 3. In Section 4, we review its computational performance. Finally, in Section 5 we discuss at some length our initial attempts to extend Multi Level Single Linkage to constrained global optimization.

2. MULTISTART

Most successful methods for global optimization involve local searches from some or all of the sample points. This presupposes the availability of some local search procedure P which starting from an arbitrary point $x \in S$, produces a local minimum x^*. Depending on what may be assumed avout f, a large number of such procedures is available from the nonlinear programming literature. We assume that P is strictly descent (Timmer [27]), such that if P is started from any point $x \in S$ and converges to a local minimum x^*, there exists a path from x to x^* along which the function values are nonincreasing. We also assume that this path is completely contained in S. Finally we assume that the number of stationary points of f, i.e. points where the gradient of f is zero, is finite.

The simplest way to make use of the local search procedure P occurs in a folklore method known as Multistart. Here, P is applied to every point in a sample, drawn from a uniform distribution over S, and the local minimum with the lowest function value found in this way is the candidate value for y_*.

An interesting analysis of Multistart was initiated in Zielinski [31] and extended in Boender and Zielinski [2], Boender and Rinnooy Kan [3], and Boender [5]. It is based on a Bayesian estimate of the number of local minima W and of the relative size of each region of attraction $\Theta_\ell = m(R(x^*))/m(S)$, $\ell = 1, \ldots, W$, where m(.) denotes Lebesque measure and a region of attraction $R(x^*)$ is defined to be the set of all points in S starting from which P will arrive at x^*.

In [5] a so-called non-informative prior distribution is specified for the unknowns W, $\Theta_1, \ldots, \Theta_W$. Given the outcome of an application of Multistart, Bayes' rule is then used to compute the posterior distribution, which incorporates both the prior beliefs and the sample information.

After lengthly calculations, surprisingly simple expressions emerge for the posterior distribution and posterior expectation of several interesting parameters (Boender [5]). For instance, if w different local minima have been found as the result of M local searches started in uniformly distributed points, then the posterior expectation of the number of local minima is

$$\frac{w(M-1)}{M-w-2} \qquad (2.1)$$

This Bayesian analysis is quite an attractive one, the more so since it can be easily extended to yield optimal Bayesian stopping rules (Boender and Rinnooy Kan [4]).

3. MULTI LEVEL SINGLE LINKAGE

In spite of the scope that Multistart offers for analysis, the procedure is lacking in efficiency. The main reason is that it will inevitably cause each local minimum to be found several times. To avoid all these time consuming local searches, P should be applied no more than once, or better still exactly once, in every region of attraction. For this purpose, the Multi Level Single Linkage method has been developed. Unlike the Single Linkage method described in Boender et al. [1], Multi Level Single Linkage focuses on the function values of the sample points to obtain an extremely simple but powerful method.

In the Multi Level Single Linkage method the local search procedure P is applied to every sample point except if there is another sample point or a previously detected local minimum within some critical distance which has a smaller function value.

Actually, the method is implemented in an iterative fashion, where points are sampled in groups of fixed size, say N. In each iteration the above rule is applied to the points of the expanded sample to determine from which sample points P should be started.

In spite of its simplicity, the theoretical properties of Multi Level Single Linkage are quite strong (Timmer [27]). If, for some $\sigma > 0$, the critical distance in iteration k is chosen to be

$$r_k = \pi^{-\frac{1}{2}}[(\Gamma(1 + \frac{n}{2})m(S)\ \sigma\ \frac{\log kN}{kN}]^{1/n}, \qquad (3.1)$$

then any local minimum x^* will be found by Multi Level Single Linkage within a finite number of iterations with probability 1. At the same time we can prove that, if $\sigma > 4$, the total number of local searches ever started by Multi Level Single Linkage is finite with probability 1 even if the sampling continues forever. In [27] it is indicated that these results are in some sense the strongest possible ones. (Actually, the above results were obtained for a slightly different version of Multi Level Single Linkage, modified to ensure that P is never applied in a point which is very close to the boundary of S or to a stationary point detected previously.)

Since Multi Level Single Linkage and Multistart result in the same set of minima with a probability that tends to 1 with increasing sample size, we can simply use the stopping rules which were designed for Multistart (Boender and Rinnooy Kan [4]).

One might believe that it is unlikely that the global minimum will be found by applying P to a sample point with a relatively high function value. It is then possible to reduce the sample by removing a certain fraction, say $1-\gamma$, of the sample points with the highest function values and to apply Multi Level Single Linkage to the reduced sample points only. Such a reduction of the sample does not significantly affect the theoretical properties of Multi Level Single Linkage. In [27] it is observed that in case of a reduction of the sample to γkN sample points in iteration k, the critical distance should still equal (3.1), but that in (2.1) M should equal γkN and not kN.

Because of the extreme simplicity of the Multi Level Single Linkage method it can be implemented very efficiently. Of course, it is not advisable to start the calculations necessary for applying the method from scratch in every iteration. Since the sample of iteration k-1 is a subset of the sample of iteration k, and since it is known in what way the critical distance varies with k, it turns out to be possible to develop an efficient dynamic implementation of the method (Timmer [27]). In this dynamic implementation, the information which is necessary to determine the starting points of the local search procedure in iteration k is determined by updating the corresponding information from iteration k-1. It turns out to be possible to implement Multi Level Single Linkage in such a way that the running time needed up to iteration k is only O(k) in expectation. Hence, the calculations needed to update the information in iteration k do not vary with the size of the complete sample, but only with the number of newly sampled points.

A fuller description of the results described in this section, including proofs, can be found in [24, 25, 27].

4. COMPUTATIONAL RESULTS

To examine the computational behaviour of Multi Level Single Linkage it has been coded in Fortran IV and run on the DEC 2060 computer of the Computer Institute Woudestein.

We tested Multi Level Single Linkage on the standard set of test functions (Dixon and Szegö [12]), which is commonly used in global optimization. These test functions are listed in Table 1.

<u>Table 1</u>

TEST FUNCTIONS

GP	Goldstein and Price
BR	Branin (RCOS)
H3	Hartman 3
H6	Hartman 6
S5	Shekel 5
S7	Shekel 7
S10	Shekel 10

In this section Multi Level Single Linkage will be compared with a few leading contenders whose computational behaviour is described in Dixon an Szegö [11]. In this reference methods are compared on the basis of two criteria: the number of function evaluations and the running time required to solve each of the seven test problems. To eliminate the influence of the different computer systems used, the running time required is measured in <u>units of standard time</u>, where one unit corresponds to the running time needed for 1000 evaluations of the S5 test function in the point (4,4,4,4).

Since both the number of function evaluations and the units of standard time required are sensitive to the peculiarities of the sample at hand, the results reported for Multi Level Single Linkage represent the average outcome of four independent runs. We applied Multi Level Single Linkage to 20% of the sample points ($\gamma = 0.2$) and choose σ to be equal to 4. After an initial sample of size 100, we increased the sample and applied Multi Level Single Linkage iteratively until the expected number of minima (2.1) exceeded the number of different minima found by less than 0.5. We did not implement the method as efficiently as possible since this is not really necessary if the sample size is moderate. Since all test functions are twice differentiable, we could use the VA10AD variable metric routine from the Harwell Subroutine Library as a local search procedure.

In Table 2 and Table 3 we summarize the computational results of Multi Level Single Linkage and compare them to those obtained for a few leading contenders as reported in Dixon and Szegö [12].

Table 2

NUMBER OF FUNCTION EVALUATIONS

METHOD	FUNCTION GP	BR	H3	H6	S5	S7	S10
Gomulka [12]	-	-	-	-	5500	5020	4860
Bremmerman [7,12]	300	160	420L	515	375L	405L	336L
Price [21]	2500	1800	2400	7600	3800	4900	4400
Törn [28,29]	2499	1558	2584	3447	3649	3606	3874
De Biase and Frontini [9]	378	597	732	807	620	788	1160
Multi Level Single Linkage	148	206	197	487	404	432*	564

L: the method did not find the global minimum.
*: the global minimum was not found in one of the four runs.

Table 3

NUMBER OF UNITS STANDARD TIME

METHOD	FUNCTION GP	BR	H3	H6	S5	S7	S10
Gomulka [12]	-	-	-	-	9	8.5	9.5
Bremmerman [7,12]	0.7	0.5	2L	3	1.5L	1.5L	2L
Price [21]	3	4	8	46	14	20	20
Törn [28,29]	4	4	8	16	10	13	15
De Biase and Frontini [9]	15	14	16	21	23	20	30
Multi Level Single Linkage	0.15	0.25	0.5	2	1	1*	2

L: the method did not find the global minimum.
*: the global minimum was not found in one of the four runs.

5. CONSTRAINED GLOBAL OPTIMIZATION

In this section we shall consider the constrained global optimization problem which is to find the global minimum x_* of an objective function f given a set of equality and inequality constraints. We assume that a set S which contains all feasible solutions (including the global one) as interior points is specified in advance. Hence, we consider the problem to find

$$y_* = \min_{x \in S} f(x) \tag{5.1}$$

$$\text{s.t. } h_i(x) = 0 \qquad i \in M_1 = \{1,\dots,m_1\}, \tag{5.2}$$

$$h_i(x) \leqq 0 \qquad i \in M_2 = \{m_1+1,\dots,m_2\}. \tag{5.3}$$

where $h_i : \mathbb{R}^n \to \mathbb{R}$ is twice continuously differentiable ($i \in M_1 \cup M_2$).

Given a point $x \in S$ it is convenient to define the following index sets

$$I(x) = \{i \in M_1 \mid h_i(x) \neq 0\}, \qquad J(x) = \{j \in M_2 \mid h_j(x) > 0\}, \tag{5.4}$$

$$\bar{I}(x) = \{i \in M_1 \mid h_i(x) = 0\}, \qquad \bar{J}(x) = \{j \in M_2 \mid h_j(x) = 0\}. \tag{5.5}$$

We define a constraint to be violated in x if it is not satified in x, i.e. if its index is in the set $I(x)$ or $J(x)$. A constraint is said to be active in x if it is violated or if its index belongs to the set $\bar{I}(x)$ or $\bar{J}(x)$. The index set of the active constraints in x is denoted by $M(x)$ and we have $M(x) = M_1 \cup J(x) \cup \bar{J}(x)$. The gradients of $f(x)$ and $h_i(x)$ ($i \in M_1 \cup M_2$) in a point x will be denoted by $\nabla f(x)$ and $\nabla h_i(x)$ respectively. $\nabla^2 f(x)$ and $\nabla^2 h_i(x)$ ($i \in M_1 \cup M_2$) will indicate the matrix of second order derivatives of f respectively h_i in x. To be able to derive necessary and sufficient conditions for a point to be a local minimum of (5.1)-(5.3), we need a so-called constraint qualification: we assume that the gradients of the active constraints, i.e. the vectors $\nabla h_i(x)$ ($i \in M(x)$) are linearly independent for every $x \in S$. Under such a constraint qualification, the Kuhn-Tucker conditions are necessary for a point x^* to be a local minimum of (5.1)-(5.3). The second order sufficient conditions for local optimality specify that, in addition to satisfying the Kuhn-Tucker conditions, the multiplier vector $\lambda_* \in \mathbb{R}^{m_2}$ (with components $\lambda_*^{(i)}$) has to satisfy

$$z^T\Big(\nabla^2 f(x^*) + \sum_{i \in M_1 \cup M_2} \lambda_*^{(i)} \nabla^2 h_i(x^*)\Big)\, z > 0 \tag{5.6}$$

for every nonzero vector $z \in \mathbb{R}^n$ satisfying $z^T \nabla h_i(x^*) = 0$ if $i \in M_1 \cup \{j \in \bar{J}(x^*) \mid \lambda_*^{(j)} > 0\}$ and $z^T \nabla h_i(x^*) \leq 0$ if $i \in \{j \in \bar{J}(x^*) \mid \lambda_*^{(j)} = 0\}$.

Only very few methods have been developed to solve the constrained global optimization problem. In [18] it is shown that the Tunneling Method can be generalized to solve problem (5.1)-(5.3). However, this approach has the same disadvantages as it has in the unconstrained case in that no guarantee of success can be provided. Some attention has been given to certain subclasses of problem (5.1)-(5.3). For instance, the case that f is concave and the constraints are convex has been described in [17, 22]. In [19] it is assumed that f is concave and that the constraints are linear. Of course, these kinds of assumptions change the structure of the problem considerably.

What happens if we try to generalize the successful method described in Section 3 to solve the constrained problem? If it would be possible to use <u>feasible</u> points only, i.e. points in S satisfying (5.2) and (5.3), then such a generalization would be straightforward. More precisely, if we draw the sample points from a uniform distribution over the feasible region and use an <u>interior point minimization technique</u> [13] as local search procedure, then we would not have to adapt our clustering procedures at all. However, it is not known how to draw points from a uniform distribution over the feasible region. Of course, one can draw the points from a uniform distribution over S and ignore the points that are not feasible. But if there are equality constraints we will never draw a feasible point, and even if M_1 is empty, such a procedure would be very inefficient for large n. Actually, even if M_1 is empty and all constraints are linear, then there is still no efficient way known to obtain points that are distributed according to a uniform distribution over the feasible region (cf. [5] for a method of generating points which are asymptotically uniform). Finally, an interior point minimization technique is not very efficient and cannot be used if there are equality constraints. We conclude that this approach does not seem very promising.

It follows that we cannot avoid infeasible points in our calculations. Recall that in the method described in Section 3, the function value of the sample points plays a crucial role: in Multi Level Single Linkage, we do not apply the local search procedure to a sample point x if another sample point is within the critical distance of x and has a smaller function value than f(x). Furthermore, one of the reasons that Multi Level Single Linkage has strong theoretical properties is that the local search procedure is descent with respect to f. If we are doomed to treat infeasible points, how should we then compare the function values of a feasible and an infeasible point? If the infeasible point has a smaller f-value this does not mean that we should not start a local search from a feasible point close to it, since the constrained local search may very well generate a sequence with increasing f-values. We conclude that we must measure the relative attractiveness of the sample points in a different way.

For this purpose we will use a penalty function. Such a penalty function measures the f-value in a point together with the extent to which it is feasible, so as to give a single value for the attractiveness of the point. Of course, it would be very helpful if we could use a penalty function ψ which has the property that there is a 1-1 correspondence between the local minima of ψ and the local minima of (5.1)-(5.3). We will call a penalty function with the above 1-1 property a doubly exact penalty function. Note that a doubly exact penalty function has stronger properties than the well known exact penalty functions [15] which are used in local constrained optimization and only have the property that all constrained local minima are also minima of the penalty function.

A doubly exact penalty function can be derived from the familiar ℓ_1-penalty function (Zangwill [30], Pietrzykowski [20])

$$\psi(x,\alpha) = f(x) + \sum_{i\in M_1} \alpha^{(i)} |h_i(x)| + \sum_{i\in M_2} \alpha^{(i)} \max(0,h_i(x)). \qquad (5.7)$$

In [20] it has been proved that for every local minimum x* of (5.1)-(5.3), there exists an $\alpha_o \in \mathbb{R}^{m_2}$ such that for all $\alpha \geq \alpha_o$, x* is a local minimum of (5.7) as well. More precisely (Charalambous [8]), if x* is a local minimum of (5.1)-(5.3) which satisfies the second order sufficient conditions with Lagrange multipliers $\lambda_*^{(i)}$ $(i \in M_1 \cup M_2)$ and if $\alpha^{(i)} > |\lambda_*^{(i)}|$, for all $i \in M_1 \cup M_2$, then x* is a local minimum of $\psi(x,\alpha)$.

It is also possible to choose α in such a way that each minimum of $\psi(x,\alpha)$ is a minimum of (5.1)-(5.3). For this purpose we introduce the following notation.

The matrix whose i-th row consists of the gradient $\nabla h_i(x)$ $(i=1,2,\ldots,m_2)$ will be denoted by A(x). $\bar{A}(x)$ is equal to A(x) except for the fact that all rows that correspond to constraints which are inactive in x are deleted. Hence, the i-th row of $\bar{A}(x)$ corresponds to the gradient of the i-th active constraints in x. If all constraints are active in x, we define a vector $\lambda(x) \in \mathbb{R}^{m_2}$ by

$$\lambda(x) = -\left(A(x)A(x)^T\right)^{-1} A(x)\nabla f(x). \qquad (5.8)$$

Note that $\left(A(x)A(x)^T\right)^{-1}$ is well defined since we assumed the gradients of the active constraints in every x to be linearly independent. If certain constraints are not active in x, then let i(j) be the j-th active constraint in x. For every active constraint in x, we now take $\lambda^{(i(j))}(x)$ to be equal to the j-th element of $\left(\bar{A}(x)\bar{A}(x)^T\right)^{-1} \bar{A}(x)\nabla f(x)$. Elements of $\lambda(x)$ corresponding to constraints that are inactive in x are taken to be equal to 0.

The function $\lambda(x) : \mathbb{R}^n \to \mathbb{R}^{m_2}$ defined in this way is bounded. To see this, let M' be an arbitrary subset of $M_1 \cup M_2$ and let S' be the largest subset of S such that for all points x in S', M' is the index set of the active constraints in x. Furthermore, let $\bar{C}(x)$ be the matrix containing the gradients of the active constraints in S', i.e. $\bar{C}(x) = \bar{A}(x)$ for all $x \in S'$. Finally, let $\bar{S}'$ be the closure of S'. It is easy to check that the constraints that are active in S' are also active in $\bar{S}'$. Since the active constraints are assumed to be independent, it follows that $\left(\bar{C}(x)\bar{C}(x)^T\right)^{-1}$ exists for all $x \in \bar{S}'$. (Note that $\bar{C}(x)$ is not equal to $\bar{A}(x)$ if $x \in \bar{S}'\ S'$.) Hence, $\left(\bar{C}(x)\bar{C}(x)^T\right)^{-1} \bar{C}(x)\nabla f(x)$ is a continuous function of x over the compact set $\bar{S}'$ and therefore it attains a maximum over this set. It follows that the supremum of $\left(\bar{C}(x)\bar{C}(x)^T\right)^{-1} \bar{C}(x)\nabla f(x)$ over S' is also bounded. Finally, since there are only a finite number of subsets M' of $M_1 \cup M_2$, and since $\bar{A}(x)$ equals $\bar{C}(x)$ in every corresponding set S', it follows that $\sup_{x\in S}(\bar{A}(x)\bar{A}(x)^T)^{-1} \bar{A}(x)\nabla f(x)$ is bounded so that a vector $\alpha \in \mathbb{R}^{m_2}$ exists which satisfies

$$\alpha^{(i)} > \sup_{x\in S} |\lambda^{(i)}(x)| \qquad (i \in M_1 \cup M_2). \qquad (5.9)$$

Given the above observation, we are now in a position to prove that, if α is chosen to satisfy (5.9) and if $x^* \in S$ is a local minimum of $\psi(x,\alpha)$, then x^* is a local minimum of (5.1)-(5.3). We can also prove that, if α satisfies (5.9) and if x^* is a local minimum of (5.1)-(5.3) which satisfies the second order sufficient conditions of (5.6), then x^* is a local minimum of $\psi(x,\alpha)$. The proofs for these results will appear elsewhere.

Thus, if α satisfies (5.9), then we almost know that ψ is a doubly exact penalty function. We only need the provision that a local minimum x^* of

(5.1)-(5.3) satisfies the second order sufficient conditions (5.6) to show that it is also a minimum of ψ. This provision is not unusual in the theory of exact penalty functions (Fletcher [14], Charalambous [8]), and actually we do not believe that it is strictly necessary.

One might argue that the requirement (5.9) prevents $\psi(x,\alpha)$ from being practically useful, since (5.9) itself defines a global optimization problem. However, this turns out to be not a serious drawback in our context; the latter global problem does not need to be solved explicitly. Moreover, we believe that a requirement like (5.9) is unavoidable in a doubly exact penalty function. We can view α as a parameter which contains global information about the problem, so that the value of ψ in a point x does not only depend on the specification of the constrained problem in x alone. Let us say that a penalty function is parameter-free if its value in any point x only depends on the properties of the constrained problem in x. We conjecture that for every continuous penalty function whose value in an arbitrary point x depends only on f(x), $h_i(x)$ $(i \in M_1 \cup M_2)$ and the values of a finite number of derivatives of f and h_i in x, there exist smooth (e.g. continuous differentiable) functions f and h_i $(i \in M_1 \cup M_2)$ such that there is no 1-1 correspondence between the minima of the penalty function and the minima of (5.1)-(5.3). The truth of this conjecture would imply that there is no parameter free doubly exact penalty function.

Through the theoretical results described above, we have laid the foundation for a stochastic method for constrained global optimization that is similar in spirit to the Multi Level Single Linkage method. The (nontrivial) details of this method do appear in [27], but we do not discuss them here. For one thing, they are beyond the scope of this contribution; for another thing, it is conceivable that there are more appropriate theoretical frameworks in which such an extension could be carried out. Even more than unconstrained global optimization, constrained global optimization represents a virgin territory that is well worth exploring in more detail.

ACKNOWLEDGEMENTS

This research of the second author was partially supported by the Netherlands Foundation for Mathematics SMC with financial aid from the Netherlands Organization for Advancement of Pure Research (ZWO). He is now associated with ORTEC consultants, Rotterdam. The first author was partially supported by a NATO Senior Scientist Fellowship.

REFERENCES

1. BOENDER, C.G.E., A.H.G. RINNOOY KAN, L. STOUGIE and G.T. TIMMER A stochastic method for global optimization, Mathematical Programming 22, (1982) 125-140.
2. BOENDER, C.G.E. and R. ZIELINSKI A sequential bayesian approach to estimating the dimension of a multinomial distribution, Technical Report, the Institute of Mathematics of the Polish Academy of Sciences (1982).
3. BOENDER, C.G.E. and A.H.G. RINNOOY KAN A Bayesian analysis of the number of cells of a multinomial distribution, The Statistician 32 (1983) 240-248.
4. BOENDER, C.G.E. and A.H.G. RINNOOY KAN Bayesian stopping rules for a class of stochastic global optimization methods, Technical Report, Erasmus University Rotterdam, The Netherlands (1983).
5. BOENDER, C.G.E. The Generalized Multinomial Distribution: A Bayesian Analysis and Application, Ph.D. Dissertation, Erasmus Universiteit Rotterdam, Centrum voor Wiskunde en Informatica, Amsterdam 1984.
6. BRANIN, F.H. and S.K. HOO A method for finding multiple extrema of a function of n variables, F.A. Lootsma (ed.), Numerical Methods of Nonlinear Optimization, Academic Press, London, 1972.
7. BREMMERMAN, H. A method of unconstrained global optimization. Mathematical Biosciences 9 (1970) 1-15.
8. CHARALAMBOUS, C. On conditions for optimality of the nonlinear ℓ_1 problem, Mathematical Programming 17 (1979) 123-135.
9. DE BIASE, L. and F. FRONTINI A stochastic method for global optimization: its structure and numerical performance [11]
10. DIXON, L.C.W. and G.P. SZEGO (eds.) Towards Global Optimization, North-Holland, Amsterdam, 1975.
11. DIXON, L.C.W. and G.P. SZEGO (eds.) Towards Global Optimization 2, North-Holland, Amsterdam, 1978.
12. DIXON, L.C.W. and G.P. SZEGO The global optimization problem. [11].
13. FIACCO, A.V. and G.P. McCORMICK Nonlinear Programming: Sequential Unconstrained Minimization Techniques, John Wiley & Sons, New York, 1968.

14. FLETCHER, R. An exact penalty function for nonlinear programming with inequalities, Mathematical Programming 5 (1973) 129-150.

15. FLETCHER, R. Penalty functions, A. Bachem et al. (eds,), Mathematical Programming, the State of Art, Springer Verlag, Berlin, 1982.

16. GOMULKA, J. Two implementations of Branin's method: numerical experience, [11].

17. HOFFMAN, K.L. (1981), A method for globally minimizing concave functions over a convex set. Mathematical Programming 20, 22-32.

18. LEVY, A. and S. GOMEZ (1980), The tunneling algorithm for the global optimization problem of constrained functions. Technical Report, Universidad National Autonoma de Mexico.

19. PATEL, N.R. and R.L. SMITH (1979), Statistical estimation of the global minimum of a concave function under linear constraints. Technical Report, University of Pittsburgh.

20. PIETRZYKOWSKI, T. An exact potential method for constrained maxima. Siam Journal on Numerical Analysis 6 (1969) 299-304.

21. PRICE, W.L. A controlled random search procedure for global optimization, [11].

22. ROSEN, J.B. Parametric global minimization for large scale problems. Technical Report, University of Minnesota, 1981.

23. RINNOOY KAN, A.H.G. and G.T. TIMMER Stochastic methods for global optimization, American Journal of Mathematical and Management Sciences (1984).

24. RINNOOY KAN, A.H.G. and G.T. TIMMER Stochastic Global Optimization Methods. Part I: Clustering Methods, Technical Report, Erasmus University Rotterdam (1985).

25. RINNOOY KAN, A.H.G. and G.T. TIMMER Stochastic Global Optimization Methods. Part II: Multilevel Methods, Technical Report, Erasmus University Rotterdam (1985).

26. RUBINSTEIN, R.Y. Simulation and the Monte Carlo Method, John Wiley & Sons, New York, 1981.

27. TIMMER, G.T. Global Optimization: A Stochastic Approach, Ph.D. Dissertation, Erasmus University Rotterdam, Centrum voor Wiskunde en Informatica, Amsterdam, 1984.

28. TORN, A.A. Cluster analysis using seed points and density determined hyperspheres with an application to global optimization. Proceeding of

the third International Conference on Pattern Recognition, Coronado, California, 1976.

29. TORN, A.A. A search clustering approach to global optimization, [11].

30. ZANGWILL, W.I. Nonlinear programming via penalty functions, Management Science 13 (1967) 344-358.

31. ZIELINSKI, R. A stochastic estimate of the structure of multi-extremal problems, Mathematical Programming 21 (1981) 348-356.

A.H.G. Rinnooy Kan and G.T. Timmer
Econometric Institute
Erasmus University Rotterdam
The Netherlands

J M SANZ-SERNA & F VADILLO

Nonlinear instability, the dynamic approach

1. INTRODUCTION

Consider an evolutionary differential equation (ordinary or partial) $u' = A(u)$, where a prime represents differentiation with respect to the time t. In the ODE case, and for each fixed t, u(t) is a d-dimensional vector and A a vector valued function of u. In the PDE case, for each fixed t, u(t) is a scalar or vector valued function of the 'space variables' x, y, ... and A an operator involving differentiation w.r. to the space variables.

Numerical methods generate approximations U_n to the true value $u(t_n)$, $t_n = nk$, which depend on the time-step parameter k. Usually the first stage in the analysis of a numerical method consists of the investigation of its convergence: does U_n tend to $u(t_n)$ as k tends to 0, n tends to ∞, $nk = t_n$ constant? It is in this analysis that the concepts of Lax-stability, Dahlquist-stability, etc... are of the outmost importance (see [11] for a survey).

Assume that the convergence of a numerical method has been established; it is still possible that for a given choice of k, or even for any such a choice, the qualitative behaviour of the numerical sequence U_0, U_1, ..., U_n, ... be completely different from that of the theoretical sequence $u(t_0)$, $u(t_1)$, ..., $u(t_n)$, ... This discrepancy which refers to n tending to ∞, k fixed cannot be ruled out by the convergence requirement, as this involves a different limit process (namely k tending to 0).

In linear, constant coefficient problems it is often possible to derive closed expressions for U_n. These can be used to derive 'stability' conditions on k (or on k and the space grid-sizes in the PDE case) which guarantee that U_n possesses the right qualitative behaviour as n increases.

In nonlinear problems it is customary to linearize the discrete equations, freeze the resulting coefficients and then demand that k satisfies all the 'stability' conditions of the resulting linear, constant coefficient problems. It has been known for a long time that this approach may fail: in 1959, Phillips [10] showed an example where U_n grew with n in spite of the fact that no growth was predicted from the linearizations. This behaviour is often referred to as nonlinear instability.

The fact that analyses based on linearization cannot accurately predict the qualitative behaviour of U_n for fixed k should not be surprising: there is a host of nonlinear phenomena (chaos, bifurcations, limit cycles ...) which cannot possibly be mimicked by a linear model.

The nonlinear instability phenomenon may be considered from a number of different points of view. A thorough survey of the literature is out of our scope here, but at least the following ideas should be briefly mentioned: (i) The role of spatial discretizations which conserve positive definite quantities in order to prevent blow ups (Arakawa [1]). (ii) The link between such discretizations and Galerkin methods (see the survey [8] by Morton). (iii) The study of time integrators which behave dissipatively in dissipative nonlinear situations (contractivity, B-stability, Dahlquist, Butcher, Spijker; see the recent book by Dekker and Verwer [5]). (iv) Fourier analysis; role of aliasing errors. (v) The important and widely quoted paper by Fornberg [6]. (vi) The connection with stability ideas in fluid mechanics investigated by Newell and his co-workers [4].

Recently there has been a growing interest in studying the fixed k behaviour from the point of view of the theory of dynamical systems [2], [7], [14]. In this paper we try to convey to numerical analysts the flavour of the powerful dynamic approach. To this end an example will be presented which illustrates several important features. The treatment of this simple case (the pendulum equation with central differences) is based on our earlier articles [12], [13], [16] and on the thesis [15]. The reader is referred to these references for applications to PDEs and for several extensions of the results discussed here.

2. THE PENDULUM EQUATION

We consider the well-known ODE system

$$u' = -\sin v, \quad v' = u, \tag{2.1}$$

describing the evolution in time of the angle v and angular velocity u of a mathematical pendulum. The phase-plane of (2.1) is displayed in many textbooks. It consists of (i) the stable equilibria $v = 2m\pi$, $u = 0$, m integer (the pendulum remains at its lowest position), (ii) the unstable equilibria $v = (2m+1)\pi$, $u = 0$, m integer (the pendulum stays at its highest position), (iii) libration orbits (the pendulum oscillates around a stable equilibrium), (iv) rotation orbits (v increases or decreases monotonically) and (v) separatrices between the libration and rotation orbits.

We also recall that in any libration domain of the (u,v)-plane (i.e. near a stable equilibrium) it is possible to change the dependent variables from (u,v) into the so-called action/angle variables (I,ϕ) [3, chap. 10]. (The abstract angle Φ is not to be confused with the physical angle v.) Among the properties of (I,ϕ) we need the following three: (i) $u = u(I,\phi)$, $v = v(I,\phi)$ are 2π-periodic with respect to ϕ, i.e. ϕ behaves like a genuine angle. (ii) I takes the value 0 at the stable equilibrium and increases away from it. (iii) In the new variables (2.1) takes the simple form

$$I' = 0, \quad \phi' = \omega(I), \tag{2.2}$$

where ω is a known function of I. It is possible to give explicit, closed expressions of the transformation $I(u,v)$, $\phi(u,v)$ and of the function $\omega(I)$ in terms of elliptic integrals, but they are not required here.

The main advantage of the new variables is that now (2.2) is readily integrated to yield

$$I(t) = I(0), \quad \phi(t) = \omega(I(0))t + \phi(0). \tag{2.3}$$

In particular it is clear that $I(u,v)$ = constant provides the equations of the orbits in the (u,v) variables.

If Δt is a given, fixed time increment, it is convenient to introduce the Δt-flow of (2.1)/(2.2). By definition this is the mapping which sends the generic point (u_o,v_o) of the phase plane into the point $(\underline{u}_o,\underline{v}_o) = (u(\Delta t),v(\Delta t))$, where $(u(t),v(t))$ is the solution of (2.1) which satisfies the initial conditions $u(0) = u_o$, $v(0) = v_o$. It follows trivially from (2.3) that in action/angle variables, the t-flow is given by the transformation $(I_o,\Phi_o) \rightarrow (\underline{I}_o,\underline{\phi}_o)$, with

$$\underline{I}_o = I_o, \quad \underline{\phi}_o = \Phi_o + \omega(I_o)\,\Delta t. \tag{2.4}$$

When (I,ϕ) are interpreted as the radius and the angle respectively in a system of polar coordinates, (2.4) is easily described: each point rotates around the origin by an angle $\omega(I_o)\Delta t$ which depends on the radius I_o. In other words, the flow leaves invariant each circle with centre at the origin. Within a circle the transformation is a rotation, but the overall effect is not that of a rigid rotation, as the angle being rotated varies with the particular circle. Transformations of this kind are called twist mappings [9].

It is useful to realize that if $\omega(I_o)\,\Delta t/2$ is a rational number p/q, then q applications of the twist (2.4) send the point back to its initial position after having completed p revolutions around the origin. On the other hand, irrational values imply that the corresponding points never return to their initial positions in the repeated iteration of the twist. In fact in this case the iterates fill densely the corresponding invariant circle and even possess a property of ergodicity: the number of iterants on any arc of the circle is proportional to the size of the arc [3].

3. LEAP FROG DISCRETIZATIONS

The system (2.1) is discretized by means of the leap frog (explicit midpoint) technique to yield, n = 1,2, ...

$$\begin{aligned} U_{n+1} &= U_{n-1} - 2k \sin V_n, \\ V_{n+1} &= V_{n-1} + 2k\, U_n. \end{aligned} \qquad (5.1)$$

Equivalently, (5.1) can be rewritten in the form

$$U_{2n} = U_{2n-2} - 2k \sin V_{2n-1}, \qquad (5.2a)$$

$$V_{2n} = V_{2n-2} + 2k\, U_{2n-1}, \qquad (5.2b)$$

$$U_{2n+1} = U_{2n-1} - 2k \sin V_{2n}, \qquad (5.2c)$$

$$V_{2n+1} = V_{2n-1} + 2k\, U_{2n}, \qquad (5.2d)$$

n = 1,2, ... , where we have simply displayed two consecutive steps of (5.1) (unrolled the loop in computer science jargon). Now (5.2) presents two remarkable features:

(i) Formulae (5.2a)/(5.2d), which compute U at an even numbered grid point and V at an odd numbered grid point, are uncoupled from formulae (5.2b)/(5.2c), which compute odd numbered Us and even numbered Vs. We refer to (5.2a)/(5.2d) as the even/odd iteration and to (5.2b)/(5.2c) as the odd/even iteration. This splitting is a result of the simple structure of (2.1) and would not carry over to more complex systems.

(ii) If U_{2n}, V_{2n+1}, U_{2n+1}, V_{2n} are regarded as approximations to p(2nk), s(2nk), r(2nk), q(2nk) respectively, where p, s, r, q are functions of t which satisfy

$$p' = -\sin s, \qquad (5.3a)$$

$$q' = r, \qquad (5.3b)$$

$$r' = -\sin q, \qquad (5.3c)$$

$$s' = p, \qquad (5.3d)$$

then (5.2) provides a consistent discretization of the system (5.3). The fact that leap frog points originating from a d-dimensional system ($d = 2$ in (2.1)) can also be regarded as consistent approximations of a 2d-dimensional system (given here by (5.3)) is universal and plays a major role in understanding leap frog discretizations. See [12] for a thorough discussion.

The system (5.3) also splits: (5.3a)/(5.3d) are not coupled to (5.3b)/(5.3c). The situation is then as follows. The values U_{2n}, V_{2n+1} originating from the even/odd iteration approximate the velocity p and position s of the pendulum (5.3a)/(5.3d), which we call the even/odd pendulum. The values U_{2n+1}, V_{2n} of the odd/even iteration approximate the velocity r and position q of a different pendulum (5.3b)/(5.3c), the odd/even pendulum. We emphasize that there is no coupling between both pendulums.

A final property of the leap frog recursion is given next.

(iii) The transformations $T_{eo}: (U_{2n-2}, V_{2n-1}) \to (U_{2n}, V_{2n+1})$, $T_{oe}: (U_{2n-1}, V_{2n-2}) \to (U_{2n+1}, V_{2n})$ preserve the area in the (u,v)-plane, i.e.whenever D is a plane domain, D, $T_{oe}(D)$, $T_{eo}(D)$ possess the same area. As in (ii), this is a general property of leap frog discretizations [12].

4. LINEARIZATIONS

Linearization of the equations of the even/odd iteration near the equilibrium $U = V = 0$ results in a recursion

$$U_{2n} = U_{2n-2} - 2k\, V_{2n-1}, \qquad V_{2n+1} = V_{2n-1} + 2k\, U_{2n},$$

whose solutions can be written in closed form in terms of the corresponding eigenvalues/vectors. It can be shown that if $k \geq 1$ then the solutions grow with n. For $k < 1$ the successive U, V values remain on an ellipse of eccentricity $(2k/(1+k))^{1/2}$ and major axis along the bisectrix of the first quadrant. Note that as k tends to 0 the orbits tend to be circles in the (u,v)-plane, thus mimicking a property of the linearized pendulum system $u' = -v$, $v' = u$. On the other hand values of k just below 1 will result in very elongated ellipses.

For the linearization of the odd/even iteration the same results hold, except that now the major axis is directed along the bisectrix of the second and fourth quadrants.

5. THE TWIST THEOREM

We have just seen that, provided that $k < 1$, the linearized theory predicts that the even/odd points will remain on an ellipse in the (u,v)-plane. The numerical experiments in the next section show that the real behaviour is far more complicated. A rigorous nonlinear analysis will now be presented. Recall that the even/odd transformation T_{eo} is an area preserving mapping which approximates consistently, i.e. up to $O(k^2)$, the 2k-flow of the even/odd pendulum. This flow is in turn a twist mapping when written in action/angle variables. The behaviour of area preserving perturbations of twist mappings is well understood. The main result is the so-called twist theorem due to Moser, see e.g. [9, p.51].

THEOREM. Assume that in (2.4) $\omega(I_o)$ is any smooth function with nonvanishing derivative and defined for $0 < a \leq I_o \leq b$. Let ε denote a positive number. Then there exists δ, depending on ω, a, b and ε but not on Δt such that any smooth area preserving mapping

$$\underline{I}_o = f(I_o,\phi_o), \qquad \underline{\phi}_o = \phi_o + g(I_o,\phi_o), \qquad a \leq I_o \leq b, \tag{5.1}$$

where f, g are 2π-periodic in ϕ_o, that is close to (2.4) in the sense that

$$|f(I_o,\phi_o) - I_o| + |g(I_o,\phi_o) - \omega(I_o)\,\Delta t| < \delta\,\Delta t, \tag{5.2}$$

possesses an invariant curve contained in $a \leq I_o \leq b$ with parametric equations

$$I_o = c + z(\xi), \qquad \phi_o = \xi + y(\xi), \tag{5.3}$$

where c is a constant and z and y are smooth 2π-periodic functions of ξ and satisfy $|z| + |y| < \varepsilon$. (Here $|.|$ denotes a suitable norm, see [9].) Furthermore, on the invariant curve (5.3) the transformation (5.1) is simply given by $\xi \to \xi+\alpha$, where $\alpha/2\pi$ is 'very irrational' (again see [9] for a precise definition). In fact each value α in the range of $\omega(I_o)\,\Delta t$, with $\alpha/2\pi$ very irrational originates such an invariant curve.

In our context $t = 2k$, (2.4) is the 2k-flow of the even/odd pendulum and (5.1) the transformation T_{eo}. The condition (5.2) is satisfied for k small enough because, by consistency, the right hand-side is $O(k^2)$. The theorem predicts that, for k small (how small depends on the distance to the equilibrium), T_{eo} possesses invariant curves which are close to the orbits of the pendulum system. On these invariants curves the transformation just acts as a rotation of irrational angle. There are also initial data which do not lie on an invariant curve. For these, the behaviour of the succesive points U_{2n}, V_{2n-1}, $n = 1, 2, \ldots$ is very involved. In any case, for k small those points must be surrounded by an invariant curve and therefore cannot scape away from the equilibrium $U = V = 0$, thus rigorously guaranteeing stability.

It is not necessary to mention that the results above also hold for the odd/even iteration.

The multidimensional analogue of the Moser twist mapping is given by the KAM (Kolmogorov-Arnold-Moser) theorem. For applications of this theorem to the analysis of numerical methods see [16].

6. NUMERICAL EXPERIMENTS

The leap frog iteration (5.1) was implemented on a computer with $U_o = 0$, V_o a parameter and U_1, V_1 taken from the application of Euler's rule. Some results will now be discussed.

In figure 1, $k = .1$ and each plotted point has coordinates U_{2n+1}, V_{2n}, so that the odd/even iteration is displayed. There are three initial conditions and for each 2,500 points were computed. For $V_o = 1, 2$ we find that the points place themselves on curves similar to the true pendulum orbits. We are thus facing the invariant curves predicted by the twist theorem. However for $V_o = 3.14$ it is quite clear that the points do not lie on a curve, but rather fill a region of finite width. Note that all three plots are elongated along the second and fourth quadrants as predicted by the linearized analysis of section 4. (A similar remark applies in subsequent experiments.)

In figure 2, k has been increased up to .625, while still displaying the odd/even iteration. There are two invariant curves corresponding to $V_o = 1, 1.5$. The value $V_o = 2$, which lead before to an invariant curve, originates now a remarkable phenomenon. The orbit consists of six suborbits so that after six iterations the computed point returns to its original suborbit. Thus, if only every sixth iterant were plotted or in other words the power T_{oe}^6 were considered rather than T_{oe}, then only one suborbit would be seen. In fact the suborbits are twist theorem invariant curves of T_{oe}^6 around a fixed point of T_{oe}^6 (or equivalently a 6-periodic point of T_{oe}).

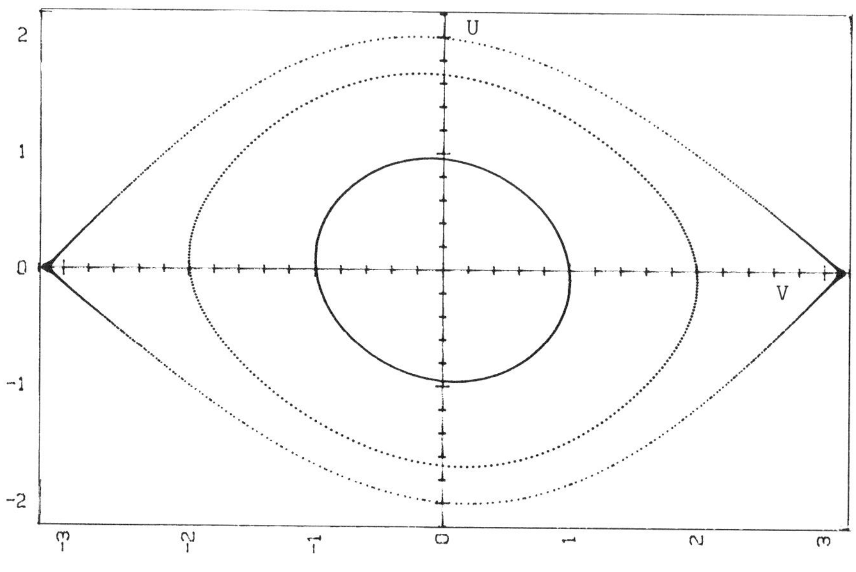

Figure 1. $k = .1$, $V_o = 1., 2., 3.14.$

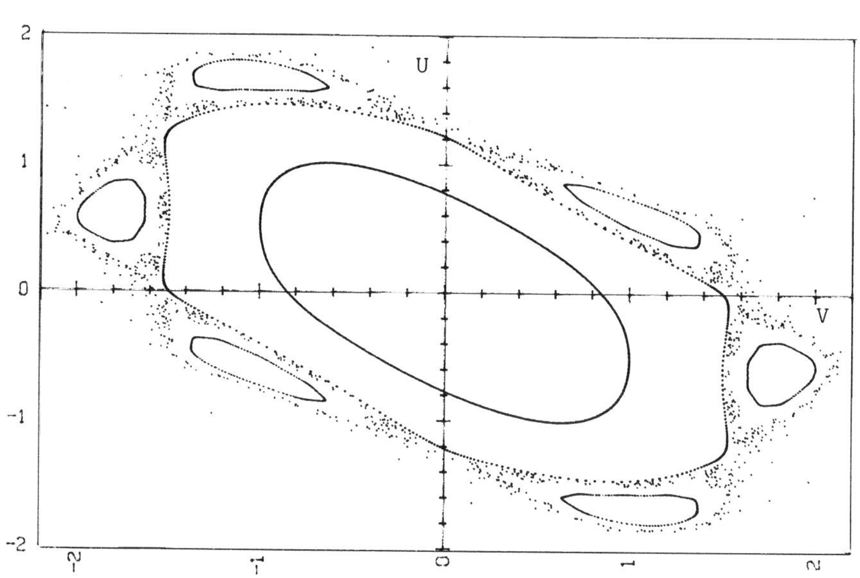

Figure 2. $k = .625$, $V_o = 1., 1.5$ (libration-like orbits), 2. (six suborbits), 2.1 (scattered points with eventual escape).

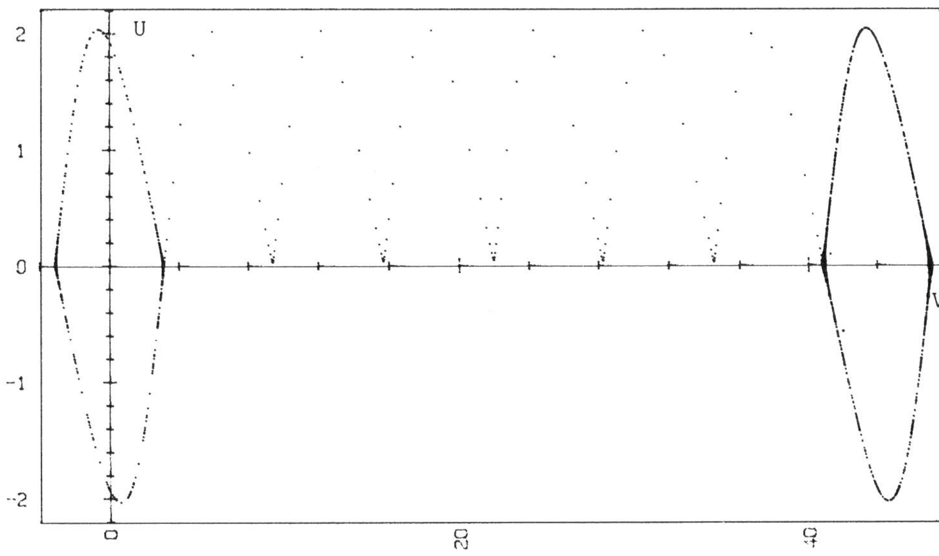

Figure 3. $k = .3$, $V_o = 3.1$. Spurious switch from libration to rotation near saddles.

Such a periodic point of T_{oe} cannot give rise to a twist invariant curve of T_{oe}, since these curves only originate from points which rotate by a very irrational angle and, of course a 6-periodic point rotates by $2p\ \pi/6$ radians.

Also displayed in figure 2 is the initial condition $V_o = 2.1$. At first, the points are scattered outside the islands corresponding to $V_o = 2.$, but eventually they escape from the plotted area. This behaviour is now possible because for the large value of k being used, the point $V_o = 2.1$, $U_o = 0$ is not surrounded by any invariant curve.

Another instance of escape is given in figure 3, where $k= .3$ and $V_o = 3.1$. There are 4,000 computed points. These first describe the expected libration orbit around $V = 0$, but later they switch to a rotation orbit and still later they settle in a libration orbit around $V = 14$. The switches occur near the unstable equilibria (saddles) $V = \pi$, $V = 13\pi$. Ushiki [14] has analyzed in detail this sort of behaviour and the reader is referred to his paper for further details. We just point out that such switches near saddles take place in an essentially un predictable way and that they are not caused by round off errors.

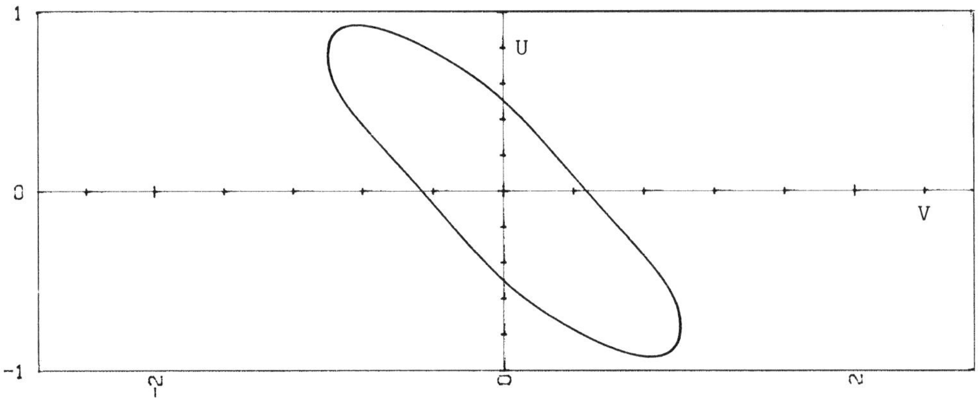

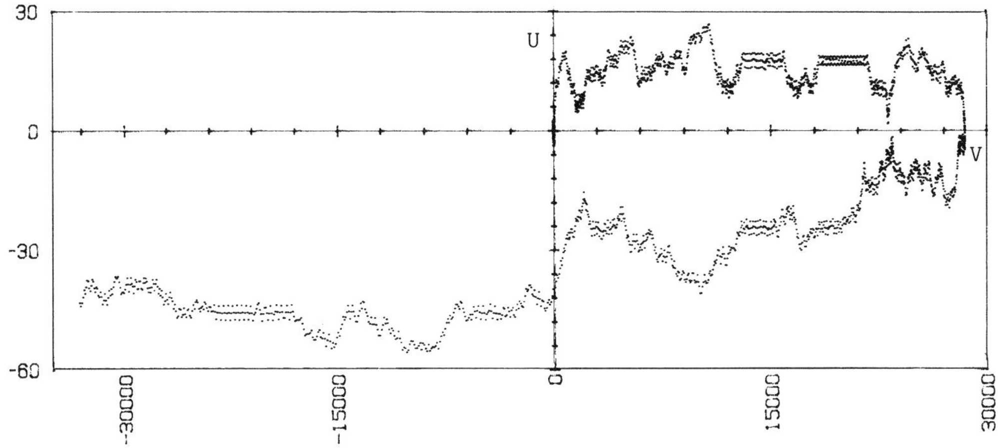

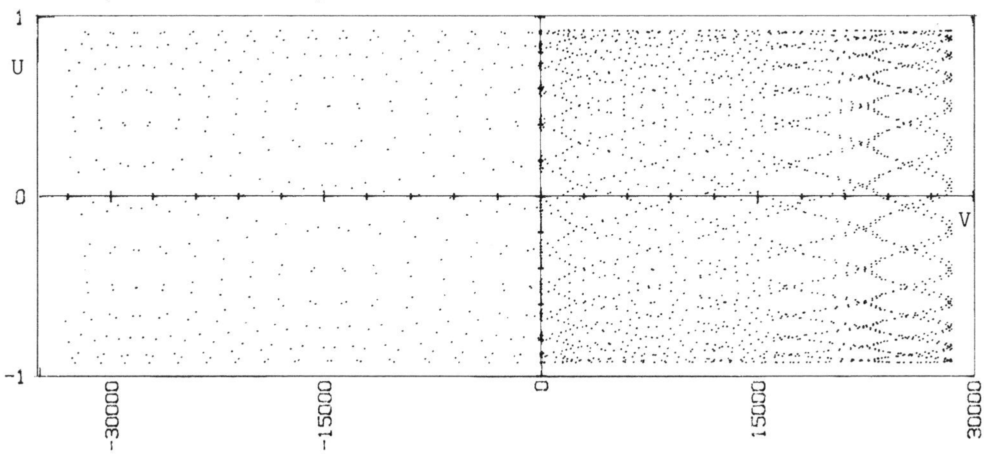

Figure 4. k = .9, V_o = 1. One numerical run mimics two pendulums. Top: odd/even points. Centre: even/odd points. Bottom: even/even points on a Lissajous-like curve.

The final experiment, in figure 4, has k= .9, near the maximum allowed by the linear condition, and $V_o = 1$. We first display the odd/even iteration which shows a familiar libration orbit with the elongation predicted by the linear theory. The even/odd iteration showed next is more interesting. (Note the change in scale of the plot.) Here we find a escape to a rotation orbit. When 2370 points have been represented, the orbit approaches a saddle, where it switches to a rotation in the opposite direction.

Comparison of the even/odd and odd/even points clearly illustrates that the leap frog points describe the motion of two uncoupled pendulums (5.3). There is no reasonable way of accounting for the behaviour of (U_n, V_n) when these are regarded, as they would normally be, as approximations to $(u(t_n), v(t_n))$, with u, v satisfying the single pendulum system (2.1). This is so even if we separate the behaviour of the odd (U_{2n+1}, V_{2n+1}) and even (U_{2n}, V_{2n}) points, a separation which has been suggested as a means for understanding the behaviour of leap frog schemes. The bottom part of figure 4 represents the even points (U_{2n}, V_{2n}). It is obvious that the dynamics displayed is unaccountable in terms of the pendulum system (2.1). Perhaps it is not without interest to point out that the plot has the appearance of a Lissajous curve. This is no surprise if we think once more of two pendulums with different frequencies.

7. CONCLUSIONS

In general, the investigation of the convergence of a numerical method provides little or no information on the qualitative behaviour of the sequence $U_0, U_1, \ldots, U_n, \ldots$ It is a (methodologically unfortunate) coincidence that in some simple, linear cases the same relations that must be imposed to achieve convergence guarantee the right qualitative behaviour. An instance is given by the familiar explicit scheme for the heat equation: the condition $r \leq \frac{1}{2}$ gives convergence and at the same time rules out the growth in time of the solution on a given grid and ensures the validity of the discrete maximum principle. (A more detailed discussion of this point has been provided by J.M. Sanz-Serna in a set of unpublished notes, available on request.)

In nonlinear situations the behaviour of $U_0, U_1, \ldots, U_n, \ldots$, when n is large, may be extraordinarily involved, as illustrated by the example considered in this paper. There is no hope of fully accounting for the phenomena involved by means of analyses based on linearization.

Fortunately, for a given, finite length of time $0 \leq t \leq T$, the use of high accuracy methods, together with small mesh-sizes guarantees that U_n is close to $u(t_n)$, thus ruling out pathological behaviours. Of course, a larger value of T will require smaller mesh-sizes.

We feel that case studies like the one presented in this paper together with estimates of the time needed for the pathologies to show up (see [16]) will play an essential role in the construction of a (Lax-like) definition of the concept of stability in nonlinear situations, a definition, in our opinion, missing at the moment. Steps in that direction have been taken within our research group and will be reported elsewhere.

ACKNOWLEDGEMENT

One of the authors (J.M.S.) has received financial support from the British Council in Spain.

REFERENCES

1. ARAKAWA, A. Computational design for long-term numerical integration of the equations of fluid motion: two-dimensional incompresible flows. Part I, J. Comput. Phys. 1 (1966) 119-143.

2. AREF, H. and DARIPA, P. K. Note on finite difference approximations to Burgers' equation, SIAM J. Sci. Stat. Comput. 5 (1984) 856-864.

3. ARNOLD, V. I. Mathematical Methods of Classical Mechanics, Springer, New York 1978.

4. BRIGGS, W. L., NEWELL, A. C. and SARIE, T. Focusing: A mechanism for instability of nonlinear difference equations, J. Comput. Phys. 51 (1983) 83-106.

5. DEKKER, K. and VERWER, J. G. Stability of Runge-Kutta Methods for Stiff Nonlinear Differential Equations, North Holland, Amsterdam 1984.

6. FORNBERG, B. On the instability of leap-frog and Crank-Nicolson approximations of a nonlinear partial differential equation, Math. Comp. 27 (1973) 45-57.

7. LAUWERIER, H. A. Dynamical Systems and Numercial Integration, Report AM-R8413, Centrum voor Wiskunde en Informatica, (Stichting Mathematical Centre), Amsterdam 1984.

8. MORTON, K. W. Initial value problems by finite difference and other methods, in The State of the Art in Numerical Analysis, D.A.H. Jacobs ed. Academic Press, London 1977.

9. MOSER, I. Stable and Random Motions in Dynamical Systems, Princeton University Press, Princeton N.J. 1973.

10. PHILLIPS, N. A. An example of nonlinear computational instability, in "The Atmosphere and the Sea in Motion", B. Bolin ed., Rockefeller Institute, New York, 1959.

11. SANZ-SERNA, J. M. Stability and convergence in Numerical Analysis I: Linear problems, a simple, comprehensive account, in Nonlinear Differential Equations and Applications, J. K. Hale and P. Martínez-Amores eds., Pitman (to appear).

12. SANZ-SERNA, J. M. Studies in numerical nonlinear instability I: Why do leapfrog schemes go unstable?, SIAM J. Sci. Stat. Comput. 6 (1985) in the press.

13. SANZ-SERNA, J. M. and VADILLO, F. Studies in numerical nonlinear instability III: Augmented Hamiltonian Systems, submitted 1985.

14. USHIKI, S. Central difference scheme and chaos, Physica 4D (1982) 407-424.

15. VADILLO, F. Inestabilidad no Lineal en Análisis Numérico: Un Estudio Dinámico, Tesis, Universidad de Valladolid, mayo 1985.

16. VADILLO, F. and SANZ-SERNA, J. M. Studies in numerical nonlinear instability II: A new look at $u_t+uu_x = 0$, submitted 1985.

J.M. Sanz-Serna
Departamento de Ecuaciones Funcionales
Facultad de Ciencias
Universidad de Valladolid
Valladolid, Spain

F. Vadillo
Departamento de Matemática Aplicada
Facultad de Ciencias
Universidad del País Vasco,
Bilbao, Spain

L N TREFETHEN

Dispersion, dissipation and stability

1. INTRODUCTION

Finite difference models are useful only when they are stable. If the von Neumann condition is violated, i.e., some Fourier mode has an amplification factor greater than 1, then the nature of the instability is obvious. But sometimes the von Neumann condition is satisfied and still the model is unstable because of the interaction of distinct Fourier modes — dispersion. The purpose of this paper is to describe such effects. A recurring theme will be the contest between dispersion, which destabilizes, and dissipation, which stabilizes but at some cost in accuracy.

This introductory section will review the basics of dispersion and dissipation, and the remaining sections will present several applications to stability.

Consider the model hyperbolic equation

$$u_t = u_x , \tag{1}$$

and the leap frog (LF) approximation

$$v_j^{n+1} = v_j^{n-1} + \lambda(v_{j+1}^n - v_{j-1}^n) , \tag{2}$$

where $\lambda = k/h = \Delta t/\Delta x$. In Fourier analysis, we insert in (2) a mode

$$v_j^n = e^{i(\omega t + \xi x)}, \qquad x = jh , \quad t = nk , \tag{3}$$

where $\xi \in [-\pi/h, \pi/h]$ is the <u>wave number</u> and $\omega \in [-\pi/k, \pi/k]$ is the <u>frequency</u>, and obtain

$$\text{LF:} \quad \sin\omega k = \lambda \sin\xi h , \tag{4}$$

the <u>dispersion relation</u> for LF. Analogously, the Crank-Nicolson (CN), Lax-Wendroff (LW), and backward Euler (BE) models of (1) have dispersion relations

$$\text{CN:} \quad 2\tan\frac{\omega k}{2} = \lambda\sin\xi h , \tag{5}$$

$$\text{LW:} \quad -i(e^{i\omega k} - 1) = \lambda\sin\xi h + 2i\lambda^2\sin^2\frac{\xi h}{2} , \tag{6}$$

$$\text{BE:} \quad -i(1 - e^{-i\omega k}) = \lambda\sin\xi h . \tag{7}$$

All of these equations approximate the ideal relationship $\omega = \xi$ for $\xi, \omega \approx 0$.

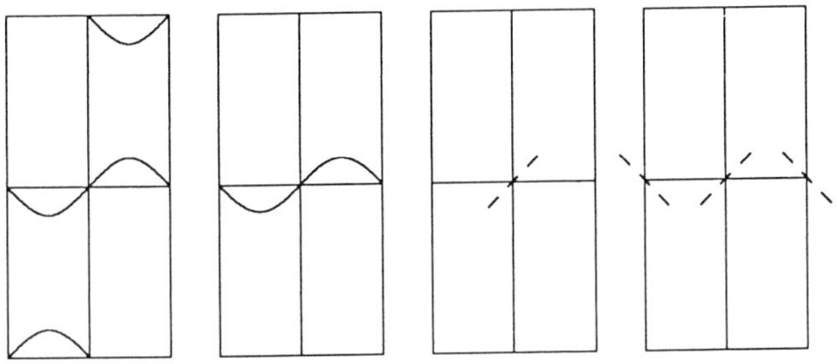

Figure 1. Dispersion relations for four models of $u_t = u_x$.

The relations (4)-(7) are plotted in Figure 1 for $\lambda = \frac{1}{2}$. In each plot the horizontal and vertical ranges are $\xi \in [-\pi/h, \pi/h]$ and $\omega \in [-\pi/k, \pi/k]$, respectively, and thus the domains are rectangles of aspect ratio 2. Alternatively, one could take ξ and ω to be arbitrary and continue the plots periodically in both directions. For LF and CN, ω is real when ξ is real, and the solid curves shown tell the whole story. For LW and BE, ω is real only for isolated values of ξ at the centers of the dashed lines, which are inclined at angles to indicate the (real) values of the derivative $d\omega/d\xi$ there. Elsewhere, ω assumes complex values that are not indicated.

Dissipation is the decay of a Fourier mode as $n \to \infty$ that comes about if $\text{Im}\omega > 0$. For a constant-coefficient model of (1), by Parseval's equality, the ℓ^2 norm of $\{v^n\}$ changes with n at a rate precisely determined by the decay of the individual Fourier components (except that the behavior of LF and other multi-level formulas is slightly more complicated). Since $\text{Im}\omega = 0$ for all $\xi \in [-\pi/h, \pi/h]$, LF and CN are nondissipative models, while since $\text{Im}\omega > 0$ for all nonzero $\xi \in [-\pi/h, \pi/h]$, LW is dissipative. BE, since it dissipates most nonzero modes but not $\xi = \pm\pi/h$, is neither dissipative nor nondissipative.

Dispersion is the interference of Fourier modes that comes about if ξ and ω are related nonlinearly — which they always are, except in unimportant special cases. To quantify this, one can consider the group velocity of a wave packet that consists locally of an oscillation $e^{i(\xi x + \omega t)}$ times a smooth envelope, defined by

$$C = -\frac{d\omega}{d\xi} . \tag{8}$$

For LF, implicit differentiation of (4) leads to the formula

$$C(\xi,\omega) = -\frac{\cos\xi h}{\cos\omega k} . \tag{9}$$

C is the velocity at which energy propagates, as can be made precise by various asymptotic arguments [11]. This is true whenever ξ , ω , and $\frac{d\omega}{d\xi}$ are all real, even though $\omega(\xi)$ may not be real at neighboring values of ξ , as with BE at $\xi = 0$ or $\xi = \pm\pi/h$ [19].

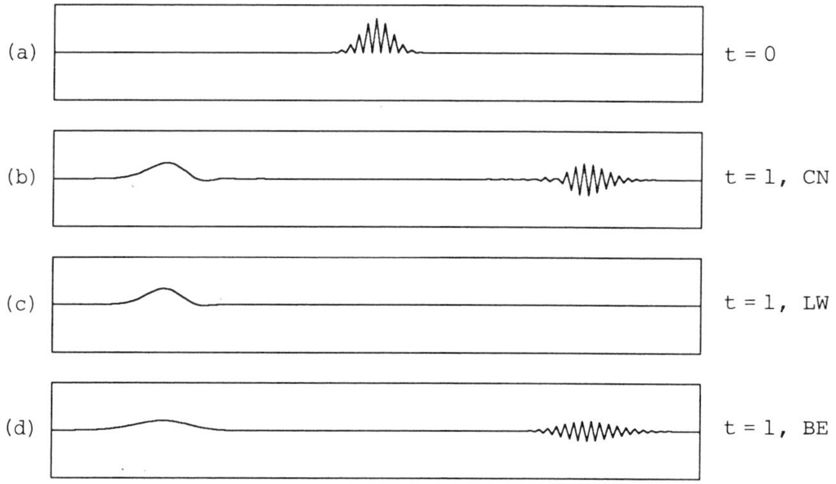

Figure 2. Separation of a wave packet into physical and parasitic components with $C \approx -1$, $C \approx +1$.

Some dispersive phenomena involve sawtoothed Fourier modes with $\xi \approx \pm\pi/h$ and/or $\omega \approx \pm\pi/k$. Figure 2 shows an example computed on the interval [0,3] with $h = .02$ and $\lambda = .5$. At $t = 0$, Figure 2a, the initial signal contains equal amounts of energy at $\xi \approx 0$ and $\xi \approx \pm\pi/h$. Figures 2b-d show the results at $t = 1$ under CN, LW, and BE. In each case about half of the energy has propagated leftward at the correct velocity $C \approx -1$. Under CN and BE, an additional spurious wave packet has also propagated rightward with $C \approx +1$. Such parasitic waves are often generated at boundaries and interfaces, and in nondissipative or weakly dissipative models it is common for them to survive and move in the wrong direction like this. See [5,16,22] for further examples.

Other dispersive phenomena involve Fourier modes with $\xi \approx \omega \approx 0$. To analyze these it is convenient to expand the numerical dispersion relation in a Taylor series in ξ about $\xi = \omega = 0$. In the Lax-Wendroff case (6), for example, one has

$$\omega = \xi\left[1 - \frac{1-\lambda^2}{6}(\xi h)^2 + \frac{i(\lambda-\lambda^3)}{8}(\xi h)^3 + \ldots\right],$$

which corresponds formally to a partial differential equation of infinite order:

$$u_t = u_x + \frac{1-\lambda^2}{6}h^2 u_{xxx} - \frac{\lambda-\lambda^3}{8}h^3 u_{xxxx} + \ldots. \qquad (10)$$

Odd-order derivatives in such an expansion introduce dispersion, and the order of the first nonzero odd derivative of order $\geqq 3$ is α, the order of dispersion of the difference model. Even-order derivatives introduce dissipation, and the order of the first one is β, the order of dissipation (possibly ∞). The order of accuracy is $\rho = \min\{\alpha,\beta\} - 1$, and is even if $\alpha < \beta$ (dispersion dominates dissipation at low wave numbers) and odd if $\alpha > \beta$ (dissipation dominates dispersion). For LF, CN, LW, and BE we have

LF: $\alpha = 3$, $\beta = \infty$, $\rho = 2$, LW: $\alpha = 3$, $\beta = 4$, $\rho = 2$,

CN: $\alpha = 3$, $\beta = \infty$, $\rho = 2$, BE: $\alpha = 3$, $\beta = 2$, $\rho = 1$.

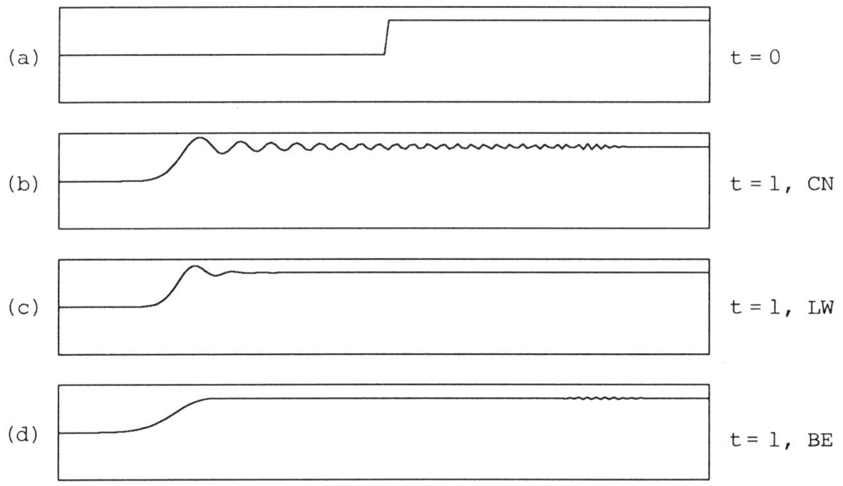

Figure 3. Oscillations around a discontinuity reveal that energy at different wave numbers travels at different group velocities.

To illustrate dispersion and dissipation for $\xi \approx \omega \approx 0$, Figure 3 repeats Figure 2 with new initial data consisting of a step function. At $t = 1$, spurious numerical oscillations have developed around the discontinuity under CN and LW. An explanation for this is that energy at various wave numbers $\xi \neq 0$ has traveled at corresponding group velocities that are different from the ideal value $C = -1$. Half-way from the origin to the discontinuity, for example, one sees a local wave number under CN corresponding to $C = -.5$. Under LW, such waves have dissipated to nearly zero amplitude, and one has to consider $C \approx -1$ to observe oscillations. Under BE, with $\beta < \alpha$, the dissipation of all modes with $C \neq -1$ is so strong that no oscillations at all are present.

The <u>modified equation</u> for a difference model is the differential equation obtained by deleting from (10) all terms except those of order 1, α, and β, i.e.

$$u_t = u_x + Ah^{\alpha-1}\frac{\partial^\alpha u}{\partial x^\alpha} + Bh^{\beta-1}\frac{\partial^\beta u}{\partial x^\beta} \tag{11}$$

for some A, B [3,23]. This equation suggests that the energy density at wave number ξ will dissipate with time approximately according to

$$\frac{\hat{v}^n(\xi)}{\hat{v}^0(\xi)} \approx \exp(-(-1)^{\frac{1}{2}\beta-1} B\xi(\xi h)^{\beta-1} t). \tag{12}$$

For an analogous estimate of dispersion, in the case of even-order schemes, we can simplify (11) by dropping the last term. From (8) we then get

$$C(\xi) \approx -1 - (-1)^{\frac{1}{2}\alpha-\frac{1}{2}} \alpha A(\xi h)^{\alpha-1}. \tag{13}$$

Though strictly applicable for dissipative models only if $\xi = 0$, this formula will make good predictions so long as ξ is small enough for the dissipation to be much weaker than the dispersion.

2. STABILITY IN ℓ^p NORMS

Usually stability is measured in the ℓ^2 norm, which is naturally related to amplification factors by Parseval's equality. But for some purposes it is useful to look at other ℓ^p norms, especially ℓ^1 and ℓ^∞, and the situation here is surprising. <u>Any finite difference model of (1) with even order of accuracy is unstable in all ℓ^p norms with $p \neq 2$</u>. This result was proved by Thomée in 1964 (unpublished), and is mentioned on p. 100 of the book by Richtmyer and Morton [14]. By the Lax equivalence theorem, it follows that convergence in ℓ^p cannot occur for arbitrary initial data, although it will still occur if the data are sufficiently smooth.

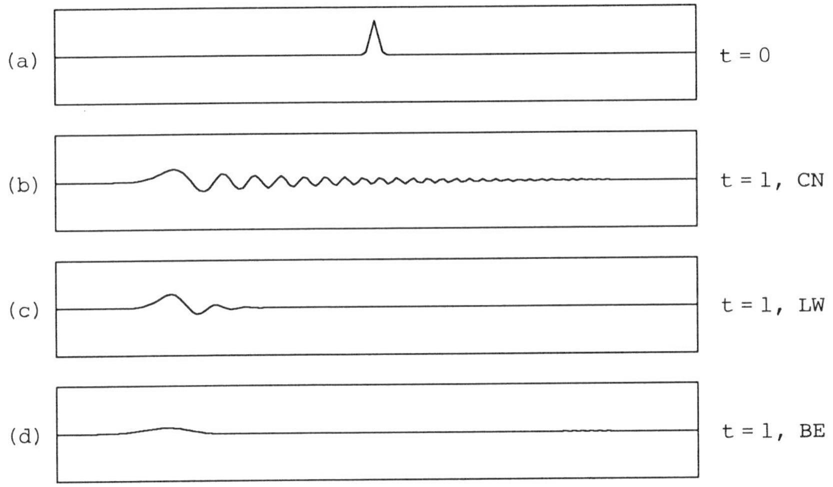

Figure 4. Dispersion of a narrow spike into a train of oscillations explains ℓ^p-instability for $p < 2$. Reversing time explains ℓ^p-instability for $p > 2$.

The explanation is dispersion. To begin with, consider a nondissipative finite difference model ($\beta = \infty$) and initial data consisting of a narrow spike of width $O(h)$ (Figure 4a). As t increases, the wave numbers will separate according to their various group velocities, as they did in Figure 3, resulting in a train of oscillations of width $O(t)$ (Figure 4b). Since $\|v^n\|_2$ is conserved in this process, clearly $\|v^n\|_p$ will increase for each $p < 2$. In fact $\|v^n\|_1$ can grow as fast as $n^{\frac{1}{2}}$, and $\|v^n\|_p$ at the rate $n^{1/p - \frac{1}{2}}$. Since $n \to \infty$ as $h \to 0$ for a fixed time t, this is a mild instability.

On the other hand, suppose we now take as data the wave train of Figure 4b, and integrate further to $t = 2$ with the sign of (1) changed — or equivalently, reverse time and go back to $t = 0$. The result will be an exact recurrence of Figure 4a. In such a process $\|v^n\|_p$ increases for $p > 2$ (up to a limited time), and the rate can be as high as $n^{\frac{1}{2} - 1/p}$. This again is a mild instability.

The same phenomenon causes ℓ^p-instability for dissipative finite difference models of even order ($\alpha < \beta$). If the order of dissipation is β, then by (12), energy at wave number ξ will attenuate on a time scale of order $t \approx h(\xi h)^{-\beta}$, i.e. $n \approx (\xi h)^{-\beta}$. Conversely, significant energy will still be present at step n for wave numbers of order $\xi h \approx n^{-1/\beta}$ and less. By (13), these wave numbers represent a range of group velocities of order $n^{(1-\alpha)/\beta}$, and therefore at step n, the train of oscillations will extend over an interval of length on the order of $(hn)n^{(1-\alpha)/\beta}$, i.e.

$$\boxed{\text{APPROXIMATE WIDTH OF REGION OF OSCILLATIONS: } hn^{\frac{\beta+1-\alpha}{\beta}}} \qquad (14)$$

(This formula predicts oscillations in Figure 3c extending over about 14 grid points, which is about right.) Now suppose the initial pulse was as narrow as possible consistent with containing only wave numbers of order at most $\xi h \approx n^{-1/\beta}$ — that is, it has width of order $hn^{1/\beta}$. Comparing this with (14) gives a broadening by a factor $n^{(\beta-\alpha)/\beta}$ from step 0 to step n, and since the ℓ^2 norm will be approximately conserved, we expect growth in $\|v^n\|_1$ by a factor of order $n^{(\beta-\alpha)/2\beta}$. More generally, for arbitrary p, with a time reversal as before to carry out the argument for $p > 2$, we get

$$\boxed{\text{APPROXIMATE UNSTABLE GROWTH RATE IN } \ell^p \text{ NORM: } n^{\frac{\beta-\alpha}{\beta}\left|\frac{1}{2}-\frac{1}{p}\right|}} \qquad (15)$$

These heuristic arguments, which were given originally in [18], reproduce the results have have been established over the years by rigorous methods of Fourier multipliers and saddle point analysis by Apelkrans, Brenner, Chin, Hedstrom, Serdyukova, Stetter, Strang, Thomée, Wahlbin, and others. In various forms the estimates (14) and (15) have been the subjects of many research papers. For (14), see for example [3,8], and for (15), see the monograph by Brenner, Thomée, and Wahlbin [2]. These authors prove that up to a constant factor, (15) gives precisely the rate of growth of powers of the discrete finite difference solution operator in ℓ^p.

The instability of (15) is weak and will rarely have much practical importance. Among our four examples, with $p = 1$ or ∞, the growth is of order $n^{1/2}$ for LF and CN, $n^{1/8}$ for LW, and n^0 (stability) for BE.

Analogous weak ℓ^p-ill-posedness occurs for hyperbolic partial differential equations in several space dimensions, where geometric focusing effects take the place of the dispersion introduced by discretization. Multidimensional finite difference models are capable of growth in ℓ^p from both sources.

3. STABILITY OF THE INITIAL BOUNDARY VALUE PROBLEM

The last section showed that ℓ^p-instability of finite difference models is caused by dispersion of wave modes with $\xi \approx \omega \approx 0$. This section will show that instability of finite difference models containing boundaries is also a matter of dispersion. This kind of instability is a more serious matter, which causes trouble frequently in practical computations. This time it is the parasitic wave modes with $\xi \approx \pm\pi/h$ and/or $\omega \approx \pm\pi/k$ that are usually responsible.

Consider (1) on the semi-infinite domain $x \geqq 0$, $t \geqq 0$. Mathematically, no boundary condition at $x = 0$ is called for, but a finite difference model will generally require one or more numerical boundary conditions. For example, each of the models LF, CN, LW, and BE requires one numerical boundary condition to determine v_0^{n+1}. The three candidates we will consider are the first-order extrapolation formulas

$$v_0^{n+1} = v_1^n, \qquad v_0^{n+1} = v_1^{n+1}, \qquad v_0^{n+1} = v_2^{n+1}. \tag{16a,b,c}$$

(The last of these is unrealistic and is included for illustration only.) The question is, which combinations of our four difference models with these three boundary conditions are stable, hence will give accurate answers?

A theory of how to answer this question was developed in the period 1960-1972. Naturally one first tries to adapt Fourier analysis, despite the bounded domain, to get a stability criterion analogous to the von Neumann test. The result is known as the <u>Godunov-Ryabenkii stability criterion</u> [14], and amounts to the following for scalar, three-point models of (1): <u>a necessary condition for stability is that the finite difference model — both interior formula and boundary condition — admit no solutions (3) with $\mathrm{Im}\omega < 0 < \mathrm{Im}\xi$</u>. It is clear why a solution of this kind causes instability: it will grow exponentially with the number of time steps. The difference from the von Neumann test is that on the one hand, an unstable mode has to satisfy two equations rather than one, while on the other, the candidates include all waves with $\mathrm{Im}\xi > 0$ rather than just $\mathrm{Im}\xi = 0$. For our twelve difference models, one can verify with a little algebra that there are no unstable modes of this type. So far, so good.

The extension of the Godunov-Ryabenkii criterion to a necessary and sufficient condition for stability was accomplished by Osher and by Kreiss and his colleagues around 1970, and is described in the well-known but difficult "GKS" paper of Gustafsson, Kreiss, and Sundström [7,12]. The essence of the Kreiss-Osher theory is the recognition that there is a second mechanism of instability to worry about: dispersive wave radiation from the boundary. For our scalar problem, the GKS stability criterion is as follows: a necessary and sufficient condition for stability is that the finite difference model admit no solutions (3) with $\text{Im}\omega \leqq 0 < \text{Im}\xi$ or with $\xi, \omega \in \mathbf{R}$ and group velocity $C \geqq 0$. "Stability" refers here to the rather complicated Definition 3.3 of [7], now commonly called "GKS-stability," but the same criterion gives the right answer for ℓ^2-stability too except in certain borderline cases (mainly $C = 0$ or $\text{Im}\omega = 0 < \text{Im}\xi$).

The GKS criterion reveals that some of our twelve difference models are unstable after all. For example, the wave

$$v_j^n = (-1)^j \qquad (\xi = \pi/h, \quad \omega = 0)$$

satisfies formulas LF, CN, and BE (not LW) and boundary condition (16c). Moreover, for each of LF, CN, and BE, Figure 1 reveals that the corresponding group velocity is $C = 1 > 0$. Therefore LF, CN, and BE are unstable with this boundary condition. To see which of the nine remaining combinations may be stable, if any, we have to check all other admissible combinations of ξ and ω also. For this simple example, that is not as hard as it sounds, and the results are listed in Table 1. Evidently four of the twelve combinations are unstable, and in one case there are two distinct unstable modes.

	LF	CN	LW	BE
(16a)	stable	stable	stable	stable
(16b)	unstable $v_j^n = (-1)^n$	stable	stable	stable
(16c)	unstable $v_j^n = (-1)^j$ or $(-1)^n$	unstable $v_j^n = (-1)^j$	stable	unstable $v_j^n = (-1)^j$

Table 1. Stable and unstable combinations of four interior formulas with three boundary conditions. For the unstable combinations, the unstable normal modes are listed.

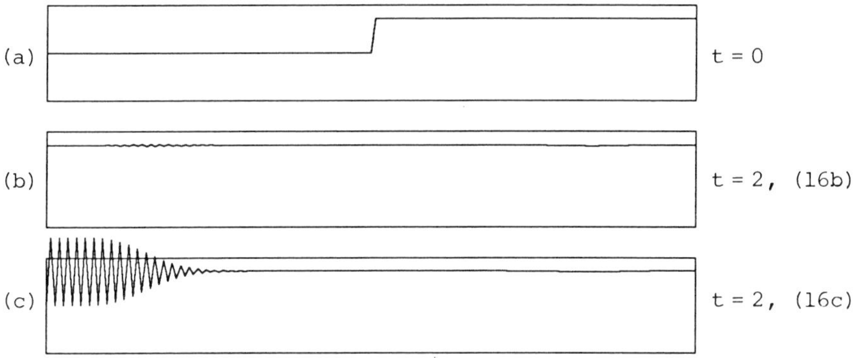

Figure 5. Two computations with formula BE — boundary condition (16b) (stable) and (16c) (unstable).

It is easy to see physically why the modes we have called unstable are troublesome. Figure 5 shows a BE computation with $h = .02$ on the usual interval [0,3] in which the initial distribution is a step function. (At the right-hand boundary the condition is $v \equiv 1$.) For the stable boundary condition (16b), the plot at $t = 2$ reveals that some oscillations have reflected from $x = 0$ into the interior with group velocity $C \approx 1$, but the solution is still approximately right. But with the unstable condition (16c), a spurious wave has been generated at $x = 0$ that will radiate rightwards into the interior forever; nor would decreasing h eliminate it. Convergence to the correct solution will not occur.

It is no coincidence that our only dissipative finite difference model, LW, turned out to be stable with all three boundary conditions. The reason is that dissipation extinguishes the parasitic waves that most often cause instability. For certain classes of problems, dissipativity actually guarantees stability — see [6]. Unfortunately, the guarantee does not extend to many problems of realistic complexity, especially when there are several space dimensions [21]. Also, the weak levels of dissipation used in many practical calculations may prevent instability in theory, but still permit behavior near enough to that of Figure 5c to be troublesome.

Further examples of stable and unstable computations can be found in [17,19,20,21]. The dispersive waves interpretation of the Kreiss-Osher theory is presented rigorously in [19]. Stability of models of initial boundary value problems is a complicated matter, however, especially when systems of equations are involved. Wave radiation from the boundary is still the general mechanism of instability, but testing for the existence of modes of this kind can be very difficult.

4. MULTIPLE BOUNDARIES AND TIME-STABILITY

In finite difference models with several boundaries, waves may bounce back and forth between them, and to understand how this affects stability we must look at reflection coefficients. To begin with, let us return to LF on $x \geqq 0$ with boundary condition (16b) of the last section. For any frequency ω, (4) and (9) (or Figure 1a) show that there are two wave numbers, a "leftgoing" value ξ_L with $C \leqq 0$ and a "rightgoing" value $\xi_R = \pi/h - \xi_L$ with $C \geqq 0$. Let us look for a constant-frequency solution of the form

$$v_j^n = e^{i\omega t}(\alpha e^{i\xi_R x} + \beta e^{i\xi_L x}), \qquad x = jh, \quad t = nk. \tag{17}$$

Inserting (17) in (16b) leads to the equation $\alpha + \beta = \alpha e^{i\xi_R h} + \beta e^{i\xi_L h}$, and since $e^{i\xi_R h} = -e^{-i\xi_L h}$ for $\xi_R = \pi/h - \xi_L$, this reduces to

$$A = \frac{\alpha}{\beta} = \frac{e^{i\xi_L h} - 1}{e^{-i\xi_L h} + 1}. \tag{18}$$

This is the <u>numerical reflection coefficient</u> for the finite difference model. Numerical experiments confirm that if a leftgoing wave at wave number ξ_L hits the boundary $x = 0$, it generates a rightgoing reflection with amplitude given by (18).

Table 1 predicted an unstable mode with $\omega = \pi/k$, $\xi_R = 0$, $\xi_L = \pi/h$. In (18) this mode gives a zero denominator and an <u>infinite reflection coefficient</u>. The existence of an infinite reflection coefficient for some ω always implies GKS-instability, but the converse does not hold, for a zero in the numerator may cancel the zero in the denominator. In fact this happens for the other unstable models of Table 1. In [19] it is shown that under reasonable assumptions, GKS-unstable models exhibit unstable growth in ℓ^2 at a rate at least $\|v^n\|_2 = O(\sqrt{n})$ in general, but at a rate at least $\|v^n\|_2 = O(n)$ if an infinite reflection coefficient is present.

In this section I want to mention two instability phenomena associated not with infinite reflection coefficients, but with <u>finite reflection coefficients greater than 1 in magnitude</u>. It should be emphasized that in a finite difference model with a single boundary, a large finite reflection coefficient (equal to 10, say) will in general not cause instability. The reason is that although a wave that hits the boundary may be amplified by reflection, this happens to it only once, so the resulting growth is bounded. If the model is consistent and GKS-stable, the amount of energy in the incident wave that excites the large reflection will decrease as $h \to 0$, and convergence to the correct solution will occur.

The first multiple-boundary instability phenomenon pertains to <u>hyperbolic equations on a bounded interval</u>. Consider the following example from a recent paper of Beam, Warming, and Yee [1] (see also [5,21]). Let (1) be modeled by BE on the usual interval $[0,3]$. At $x=3$ we impose the boundary condition $v_J^n = 0$, where $J = 3/h$. Inserting the ansatz (17) with x replaced by $x-3$ gives a corresponding reflection coefficient

$$A_R = \frac{\beta}{\alpha} = -1, \tag{19}$$

independent of ω. At $x=0$ we consider both boundary conditions (16a) and (16b). The reflection coefficients $A_L = \alpha/\beta$ are

$$(16a):\ A_L = \frac{e^{i\xi_L h} - e^{i\omega k}}{e^{-i\xi_L h} + e^{i\omega k}}, \qquad (16b):\ A_L = \frac{e^{i\xi_L h} - 1}{e^{-i\xi_L h} + 1}. \tag{20a,b}$$

Neither of these denominators is ever zero for waves admitted by BE (unlike LF), so each boundary is individually GKS-stable and has finite reflection coefficients. By Theorem 5.4 of [7], it follows that the two-boundary model is GKS-stable too.

Nevertheless, only (16b) is usable in practice for large λ. The attraction of an implicit formula like BE is its applicability to simulations of the steady-state limit $t \to \infty$, because according to von Neumann analysis, the time step can be taken arbitrarily large without causing instability. But Beam, et al. found that when BE is applied with J even and λ large to (1), or to a more realistic problem in transonic gas dynamics, the solutions with (16b) converge as $t \to \infty$ but those with (16a) grow exponentially in t. To illustrate this, Figure 6 shows the solution obtained under BE and (16a) with initial data as in Figure 4, but computed with the large Courant number $\lambda = 300$ ($k = 6$) and plotted at large time intervals $\Delta t = 24$ (4 times steps each). At first, the energy nearly dies away, as it should for the mathematical problem. But a small sawtoothed signal remains that begins to grow unboundedly. Obviously, the steady-state solution $v \equiv 0$ will not be approached as $t \to \infty$.

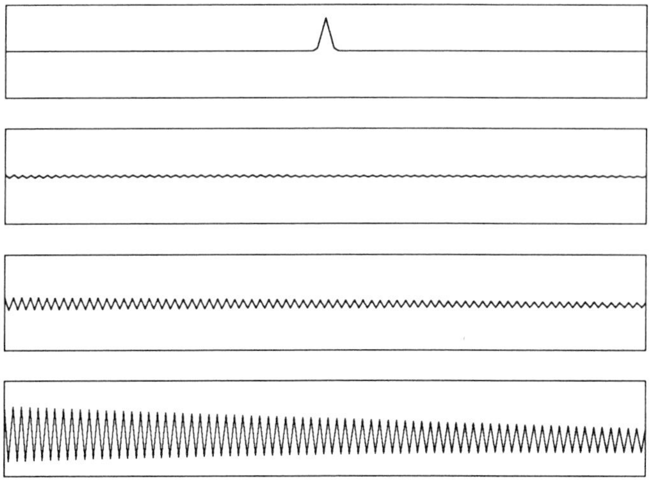

Figure 6. Exponential growth in t caused by repeated reflection between boundaries with a reflection coefficient greater than 1 .

This behavior can be explained by reflection coefficients. For $\lambda \gg 1$, inserting (3) in (7) gives

$$\xi_L = O(1/\lambda) , \qquad \xi_R = \pi/h + O(1/\lambda) ,$$

independently of ω . In particular, these formulas hold for $\omega = \pi/k$. But for that frequency, (20a) gives a large reflection coefficient

$$A_L = O(\lambda) . \qquad (21)$$

Now from (19) and (21), we see that a wave may potentially bounce back and forth between the two boundaries, increasing in amplitude by a factor $O(\lambda)$ with each circuit. This will cause growth in any norm at a rate const^t . In Figure 6d, the solution shown is very nearly an eigenmode of the form (17) of the finite difference model with two boundaries. The sawtoothed component, which dominates, is a rightgoing wave that dissipates somewhat as it travels from $x = 0$ to $x = 3$. The fact that the sawtooth is centered below the axis reveals that there is also a nonzero smooth leftgoing component predicted by (19).

By contrast, it can be shown that $\mathrm{Re}(e^{i\xi_L h}) \geq 0$ for all ω with $\mathrm{Im}\omega \leq 0$

under BE, and so (20b) implies $|A_L| \leq 1$. Together with (19), this shows that the two-boundary model with boundary condition (16b) does not admit any solutions consisting of waves that bounce back and forth and increase in amplitude.

Both of these finite difference models are GKS-stable, and will converge to the correct solution as $h \to 0$ for any fixed t. However, the model based on (16a) will fail to converge to anything for fixed h, k as $t \to \infty$. It may also give poor results for finite t with the nonzero values of h used in practice. A finite difference model that admits no exponentially growing solutions as $t \to \infty$ is said to exhibit time-stability (= "practical stability" = "P-stability"). We have illustrated that at least in some cases, time-stability is determined by numerical reflection coefficients.

These arguments can be generalized. Beam, Warming, and Yee consider not just BE but any A-stable time-discretization (including CN) of the usual 3-point difference operator in space, and also space and space-time extrapolation boundary conditions of higher order. For a rigorous analysis of general two-boundary problems by reflection coefficients, see [21]. It is proved there that any two-boundary model with reflection coefficients satisfying $\|A_L\| \, \|A_R\| < 1$ must be time-stable. It is also proved that in the case of a dissipative model, it is enough for an estimate analogous to $\|A_L\| \, \|A_R\| < 1$ to hold for the differential equation. We write norms rather than absolute values for these results since in general, a hyperbolic system or a finite difference model admits several leftgoing and rightgoing modes, and the reflection coefficients imposed by the boundary conditions then give way to reflection matrices.

The second multiple-boundary instability phenomenon of this section is a more speculative idea that I want to mention quite briefly, having to do with hyperbolic equations in a two-dimensional domain with a corner. Suppose a hyperbolic equation is given on $x, y, t \geq 0$ subject to boundary conditions along $x = 0$ and $y = 0$, each of which would yield a well-posed problem on a half-plane. (The criterion for well-posedness on a half-plane, due to Kreiss in 1970 [10], is closely analogous to the GKS criterion for stability of a difference model on an interval: the general mechanism of ill-posedness is wave radiation from the boundary, and a problem is always ill-posed if there exists an infinite reflection coefficient at some frequency [9].) Will the problem with the corner be well-posed? This question was first considered by

Osher, and he gave several examples showing that in general the answer is no [13]. The reason is that in some circumstances, a trapped wave packet may bounce back and forth between the two boundaries near the corner, and if its amplitude increases by some finite factor greater than 1 with each reflection, the result will be exponential growth. This may sound like "practical ill-posedness" rather than true ill-posedness. But what makes this problem interesting is that since the wave packet might be arbitrarily close to the corner, the travel time between reflections might be arbitrarily small, and therefore the time constant of such exponential growth cannot be bounded.

In a beautiful paper in 1975, Sarason and Smoller described how to analyze problems like this by invetigating propagation of wave packets along rays [15]. For each frequency ω, the dispersion relation (or characteristic variety) for a differential equation or difference model consists of one or more curves in the wave number space (ξ_x, ξ_y). When a wave packet with parameters (ξ_x, ξ_y, ω) hits the boundary $x = 0$, ξ_y will remain constant but ξ_x may jump to another position on this curve, and likewise with the roles of x and y reversed at $y = 0$. On the other hand the velocity of propagation of the wave packet in space is given by the vector group velocity $C = -\nabla_\xi \omega$. Given the plot of a dispersion relation, one can apply these considerations to see readily whether the problem in the corner is susceptible to modes that bounce back and forth endlessly between the two boundaries, in which case ill-posedness will occur if the boundary conditions happen to impose large reflection coefficients. On the other hand if all wave packets eventually escape to infinity, the problem is well-posed regardless of the boundary conditions.

Sarason and Smoller showed that for any 2×2 strictly hyperbolic problem — such as the second-order wave equation — the dispersion curves are simply ellipses, leading to the conclusion that all wave packets eventually escape to infinity, so well-posedness is assured. But for finite-difference models, the situation is different, for the dispersion plots are more complicated (see Figure 7 of [16]). It seems certain that one could devise a finite difference model of the wave equation in a corner that was unstable because of repeated reflection at the boundaries. The growth rate would be a catastrophic const^n, despite the well-posedness of the underlying differential equation and the finiteness of all reflection coefficients. This increase in severity from const^t to const^n would be a direct result of the distance between the boundaries becoming vanishingly small at the corner.

It is unlikely that precisely this kind of corner instability ever appears in models of practical interest. But analogous one-dimensional instabilities involving reflections between close together boundaries do occur — see Section 3 of [21].

5. VARIABLE COEFFICIENTS AND NONLINEARITY

In this final section I will mention some ways in which dispersion and dissipation enter into stability questions for problems with variable coefficients or nonlinearity.

The propagation of wave packets under a finite difference model of a linear problem with variable coefficients, such as

$$u_t = a(x)u_x , \tag{22}$$

has been investigated by methods of geometrical optics by Giles and Thompkins [5]. In this situation, even if there is no dissipation, both the amplitude and the wave number of a wave packet change continually in accordance with formulas that depend on the derivative $\frac{da(x)}{dx}$. These changes are numerical artifacts, little related to the behavior of the problem being modeled, and energy is in general not conserved. Giles has made a movie that illustrates the effects of variable coefficients compellingly. In one of its demonstrations, (22) is modeled on an unbounded domain by a nondissipative formula with a variable coefficient $a(x) \leqq a_0 < 0$. A numerical wave packet is then observed to oscillate left and right forever between two extreme positions, alternating between smooth and sawtoothed form with each change in direction, even though the differential equation contains no boundaries and admits rightward propagation only.

As an application of these ideas to stability, Giles and Thompkins consider the implications for time-stability of finite difference models with multiple boundaries. In some cases at least, the analysis of the last section can be adapted in a straightforward way. Now, instead of looking for a reflection coefficient bound $||A_L|| \, ||A_R|| < 1$ to ensure stability, one requires this product to be less than an appropriate frequency-dependent constant. The details are given in [5].

The same principles that govern smoothly varying coefficients also apply to models with smoothly varying mesh spacings. For mesh refinements of a discontinuous sort, on the other hand, it is more appropriate to treat each jump as an interface and look at reflection and transmission coefficients [21].

Another connection of dispersion and dissipation to stability of problems with variable coefficients concerns differential equations on an unbounded domain. When can ℓ^2-stability of a finite difference model of a symmetric hyperbolic system with variable coefficients be inferred from stability of the "frozen" problems for each x ? This question, of obvious practical importance, received much attention in the 1960's, and is discussed in Chapter 5 of the book by Richtmyer and Morton [14]. One result, the Lax-Nirenberg theorem, guarantees stability so long as the coefficients are twice continuously differentiable and the amplification matrix has norm bounded by 1 for each x . A different one, due to Kreiss, guarantees stability when the coefficients are merely Lipschitz continuous provided that the difference model is dissipative and of odd order — that is, $\beta < \alpha$, the same as the condition for ℓ^p-stability in Section 2. Parlett later showed that $\beta \leq \alpha+1$ is good enough in the case of a strictly hyperbolic system such as (22). It would be interesting to know whether the difference between the Lax-Nirenberg and Kreiss-Parlett results can be explained physically in terms of dispersive and dissipative propagation of waves.

When we turn to nonlinear problems, very little is known in a general way about stability. It seems likely that there may be nonlinear stability principles for initial boundary value problems, analogous to the Kreiss-Osher theory of Section 3 for the linear case, that could be obtained by consideration of nonlinear dispersion relations [24], but this has not been investigated. What has been looked at is chiefly the example of the inviscid Burgers equation,

$$u_t = uu_x = (\tfrac{1}{2}u^2)_x , \tag{23}$$

modeled by a leap frog discretization of the form

$$\frac{v_j^{n+1} - v_j^{n-1}}{2k} = (1-\theta)\, v_j^{n+1} \left(\frac{v_{j+1}^n - v_{j-1}^n}{2h}\right) + \theta\left(\frac{(\tfrac{1}{2}v_{j+1}^n)^2 - (\tfrac{1}{2}v_{j-1}^n)^2}{2h}\right) \tag{24}$$

for some $\theta \in [0,1]$. The question of the behavior of (24) dates to Phillips in 1959, and has been studied since then by Stetter, Arakawa, Richtmyer and Morton, Fornberg [4], Kreiss and Oliger, Majda and Osher, Newell, and Briggs, Newell, and Sarie [25]. The omitted references can be found in the papers cited.

For accurate simulation of shock speeds, one might expect that one should take $\theta = 1$, since the model is then in conservation form. But it turns out that $\theta = 2/3$ is also a critical value, at which the energy would be conserved if the time discretization were exact. The papers mentioned show that for $\theta > 2/3$, a 3h-periodic mode of the form $(\ldots, -c, 0, c, -c, 0, c, \ldots)$ with $c > 0$ will blow up super-exponentially under (24), while for $\theta < 2/3$, the same is true of a mode $(\ldots, c, 0, -c, c, 0, -c, \ldots)$. Therefore (24) is unstable for all $\theta \neq 2/3$. In computational experiments, however, the explosion is usually observed only in the latter case. The reason is that for $\theta > 2/3$, local patterns approximating the configuration $-c, 0, c$ tend to disperse into waves of constant amplitude radiating in both directions. By contrast, with $\theta < 2/3$, irregular initial oscillations tend to concentrate into the unstable configuration and blow up quickly. (The same also occurs with $\theta > 2/3$ if a boundary with a homogeneous boundary condition is introduced.) Thus (24) seems to be an instance in which dispersion is not the fundamental mechanism of instability, yet it is the factor which controls whether that mechanism is excited.

Finally, consider $\theta = 2/3$ and values of u bounded initially above 0. Either of these conditions is enough to make (24) appear stable in most experiments for short times. Nevertheless, Briggs, et al. discovered that when very many time steps are taken, a fascinating new kind of instability occurs [25]. An example is shown in Figure 7, in which (24) has been applied with $\lambda = \frac{1}{2}$ on the usual grid with periodic boundary conditions. The initial signal is 4h-periodic, except that a random perturbation has been added of amplitude about .01. At $t = 12$, nothing surprising has happened. But at $t = 24$, the signal has begun to undulate in a regular way — an effect one would never encounter in a linear situation. By $t = 36$, the undulations have become dangerously large, and in a few more time steps the CFL limit will be exceeded locally and an explosion will take place. Briggs, et al. show that what is going on is a systematic process of <u>nonlinear focusing</u> of energy, and that this will always occur on any sufficiently fine mesh. The precise positions at which large amplitudes first appear are more or less random, but the nature of their appearance and growth follows a predictable pattern. We conclude that (24) is unstable even for $\theta = 2/3$ and with initial data of one sign.

The lesson of this last example, as of all of the examples of this paper, is that instabilities may have subtle explanations. But they always <u>do</u> have explanations — often related to dispersion of waves.

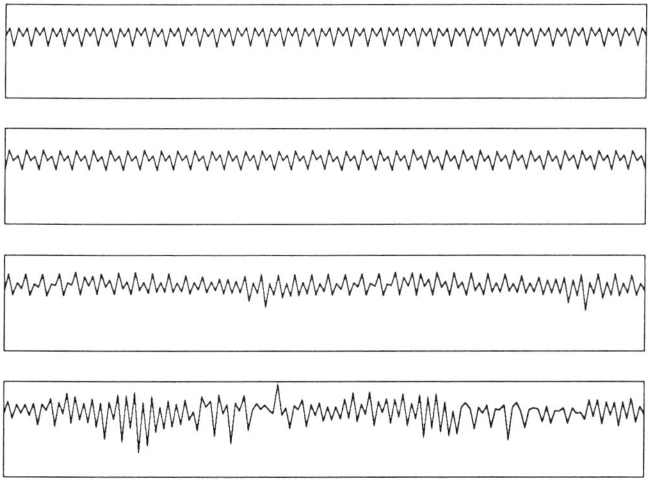

Figure 7. Instability brought about by nonlinear wave focusing in a discrete model of the Burgers equation (after Briggs, Newell, and Sarie).

REFERENCES

1. BEAM, R.M., WARMING, R.F., and YEE, H.C. Stability analysis of numerical boundary conditions and implicit difference approximations for hyperbolic equations, J. Comp. Phys. 48 (1982) 200-222.

2. BRENNER, P., THOMÉE, V., and WAHLBIN, L. Besov Spaces and Applications to Difference Methods for Initial Value Problems, Springer Lect. Notes in Math. v. 434, New York, 1975.

3. CHIN, R.C.Y. and HEDSTROM, G.W. A dispersion analysis for difference schemes: tables of generalized Airy functions, Math. Comp. 32 (1978) 1163-1170.

4. FORNBERG, B. On the instability of Leap-Frog and Crank-Nicolson approximations of a nonlinear partial differential equation, Math. Comp. 27 (1973) 45-57.

5. GILES, M.B. and THOMPKINS, W.T., Jr. Propagation and stability of wave-like solutions of finite difference equations with variable coefficients, J. Comp. Phys., to appear.

6. GOLDBERG, M. and TADMOR, E. Scheme-independent stability criteria for difference approximations of hyperbolic initial-boundary value problems. II, Math. Comp. 36 (1981) 603-626.

7. GUSTAFSSON, B., KREISS, H.-O., and SUNDSTRÖM, A. Stability theory of difference approximations for initial boundary value problems. II, Math. Comp. 26 (1972) 649-686.

8. HEDSTROM, G.W. The rate of convergence of some difference schemes, SIAM J. Numer. Anal. 5 (1968) 363-406.

9. HIGDON, R.L. Initial-boundary value problems for linear hyperbolic systems, SIAM Review, to appear.

10. KREISS, H.-O. Initial boundary value problems for hyperbolic systems, Comm. Pure Appl. Math. 23 (1970) 277-298.

11. LIGHTHILL, J. Waves in Fluids, Cambridge University Press, 1978.

12. OSHER, S. Systems of difference equations with general homogeneous boundary conditions, Trans. Amer. Math. Soc. 137 (1969) 177-201.

13. OSHER, S. Hyperbolic equations in regions with characteristic boundaries or with corners, in Numerical Solution of Partial Differential Equations III, B. Hubbard, ed., Academic Press, New York, 1976.

14. RICHTMYER, R.D. and MORTON, K.W. Difference Methods for Initial-Value Problems, Wiley-Interscience, New York, 1967.

15. SARASON, L. and SMOLLER, J.A. Geometrical Optics and the corner problem, Arch. Rat. Mech. Anal. 56 (1975), 34-69.

16. TREFETHEN, L.N. Group velocity in finite difference schemes, SIAM Review 24 (1982) 113-136.

17. TREFETHEN, L.N. Group velocity interpretation of the stability theory of Gustafsson, Kreiss, and Sundström, J. Comp. Phys. 49 (1983) 199-217.

18. TREFETHEN, L.N. On ℓ^p-instability and oscillation at discontinuities in finite difference schemes, in Advances in Computer Methods for Partial Differential Equations V, R. Vichnevetsky and R.S. Stepleman, eds., IMACS, 1984.

19. TREFETHEN, L.N. Instability of difference models for hyperbolic initial boundary value problems, Comm. Pure Appl. Math. 37 (1984) 329-367.

20. TREFETHEN, L.N. Stability of hyperbolic finite-difference models with one or two boundaries, Proc. AMS-SIAM Summer Seminar on Large-Scale Computations in Fluid Mechanics, AMS, Providence, RI, to appear.

21. TREFETHEN, L.N. Stability of finite difference models containing two boundaries or interfaces, Math. Comp., to appear.

22. VICHNEVETSKY, R. and BOWLES, J.B. Fourier Analysis of Numerical Approximations of Hyperbolic Equations, SIAM, Philadelphia, 1982.

23. WARMING, R.F. and HYETT, B.J. The modified equation approach to the stability and accuracy analysis of finite-difference methods, J. Comp. Phys. 14 (1974) 159-179.

24. WHITHAM, G.B. Linear and Nonlinear Waves, Wiley-Interscience, New York, 1974.

25. BRIGGS, W.L., NEWELL, A.C., and SARIE, T. Focusing: a mechanism for instability of nonlinear finite difference equations, J. Comp. Phys. 51 (1983) 83-106.

L. N. Trefethen
Department of Mathematics
Massachusetts Institute of Technology
Cambridge, Massachusetts 02139
USA

J G VERWER

Convergence and order reduction of diagonally implicit Runge–Kutta schemes in the method of lines

1. INTRODUCTION

The method of lines (MOL) idea is simple in concept: for a given time dependent partial differential equation (PDE) discretize the space variables so that the equation is converted into a continuous time system of ordinary differential equations (ODEs). This ODE system is then numerically integrated by an integration scheme, often one which can handle *stiffness.* Various known numerical schemes for PDEs can be viewed in this way. This contribution is devoted to an analysis for the *full error* of implicit Runge-Kutta MOL schemes. We will particularly concern ourselves with a class consisting of four known *diagonally implicit methods* although much of this paper will apply to other schemes as well. However, within the class of general implicit methods there is a significant computational advantage in diagonally implicit RK (DIRK) methods, especially for PDEs. With the exception of special circumstances, other types of implicit RK methods are in fact of rather limited practical value here.

An overview of the paper reads as follows. In §2 we discuss the type of evolution problems our analysis applies to. The third paragraph is devoted to preliminaries on the discretization. Here we present the four DIRK schemes and we anticipate on the *convergence analysis* which is presented for these schemes in detail in §4. This analysis is centered around the semi-discrete approximation, i.e., the ODE system. That means that the stability concept we use is borrowed from the field of nonlinear, stiff ODEs [7]. Our error analysis is reminiscent of the analysis developed in the B-convergence theory by Frank, Schneid & Ueberhuber [9,10]. The central theme of this theory is that of *order reduction.* We examine this unwanted phenomenon in detail for a 3-rd and 4-th order DIRK scheme in the MOL framework. An interesting feature of these DIRK schemes is that the reduction for the global error is less than for the local error, although it still may be considerable when it occurs. To illustrate that the results of our analysis have real practical significance we have performed a number of numerical experiments which are presented in §5. There we also summarize some conclusions on the merits of higher order DIRK schemes in the method of lines.

2. PRELIMINARIES ON THE PROBLEM CLASS

We consider a real abstract Cauchy problem

$$u_t = \mathcal{F}(x,t,u), \quad 0 < t \leqslant T, \quad u(x,0) = u^0(x), \tag{2.1}$$

where $\mathcal{F}$ represents a partial differential operator which differentiates the unknown function $u(x,t)$ w.r. to its space variable x in the space domain in $\mathbb{R}, \mathbb{R}^2$ or $\mathbb{R}^3$. $\mathcal{F}$ should not differentiate w.r. to the time variable t. The function $u(x,t)$ may be a vector function. Boundary conditions are supposed to be included in the definition of $\mathcal{F}$.

To the problem (2.1) we associate a real Cauchy problem for an ODE system,

$$\dot{U} = F(t,U), \quad 0 < t \leqslant T, \quad U(0) = U^0, \; F(t,\cdot):\mathbb{R}^m \to \mathbb{R}^m, \tag{2.2}$$

which is defined by a discretization of the space variable in (2.1). For the moment it is not necessary to discuss in detail how the semi discrete, continuous time approximation (2.2) arises from (2.1). Nor is it necessary, for the time being, to be specific about the partial differential equation. The reason is that our convergence analysis is centered around the ODE system (2.2). This is most convenient for the analysis and allows for the general treatment we aim at. We merely assume that U and F represent the values of grid functions on a space grid covering the space domain of (2.1). Further, we let h refer to the grid spacing, i.e., to the grid distances which may vary over the grid. In what follows, $h \to 0$ means that the grid is refined arbitrary far in a suitable manner. Note that the dimension m of problem (2.2) depends on h. The formulation (2.2) of the semi-discrete problem indicates that we concentrate on finite difference space discretizations. However, finite element or spectral methods could also be considered.

Let $\|\cdot\|$ be a vector norm on $\mathbb{R}^m$ (we shall use the same symbol for the subordinate matrix norm) and $\mu\,[\cdot]$ the corresponding logarithmic matrix norm. Let $F'(t,\cdot)$ be the Jacobian matrix of $F(t,\cdot)$. Our analysis applies to problems (2.1)- (2.2) for which $\mu[F'(\cdot,\zeta)]$, $\zeta \in \mathbb{R}^m$ can be bounded from above by a constant, $\mu_{\max}$ say, which is *independent of the grid spacing*, i.e., $\mu_{\max}$ should satisfy

$$\mu_{\max} \geqslant \max_{\zeta \in \mathbb{R}^m} \mu[F'(\cdot,\zeta)] = \max_{\zeta \in \mathbb{R}^m} \lim_{\Delta \downarrow 0} \frac{\|I + \Delta F'(\cdot,\zeta)\| - 1}{\Delta} \tag{2.3}$$

uniformly in h. We let ζ lie in the whole of $\mathbb{R}^m$ for convenience of presentation. In actual applications it suffices to take ζ in a tube around the exact solution. For inner product norms $\|\zeta\| = (<\zeta,\zeta>)^{1/2}$ condition (2.3) can be reformulated as the one- sided Lipschitz condition (see [7], §1.5)

$$<F(\cdot,\tilde{\zeta}) - F(\cdot,\zeta), \tilde{\zeta} - \zeta> \leqslant \mu_{\max}\|\tilde{\zeta} - \zeta\|^2, \quad \forall \tilde{\zeta},\zeta \in \mathbb{R}^m. \tag{2.4}$$

Hypothesis (2.3), or (2.4), implies that any two solutions $\tilde{U}$, U of (2.2) satisfy the *exponential stability estimate* (a result due to Dahlquist [6])

$$\|\tilde{U}(t) - U(t)\| \leqslant e^{\mu_{\max} t}\|\tilde{U}(0) - U(0)\|, \quad \forall t \in [o,T], \tag{2.5}$$

uniformly in h. Hence, in view of this well-posedness inequality, conditions (2.3)-(2.4) are natural. We wish to remark, however, that given a certain pair of problems (2.1)-(2.2), it may be far from trivial to select a specific norm for which (2.3) or (2.4) can proved to be valid.

Example 2.1. To illustrate the foregoing we mention two equations which were analysed in [18]. The first is the scalar, nonlinear parabolic equation

$$u_t = f(t,x,u, \frac{\partial}{\partial x}(d(x,t)\frac{\partial u}{\partial x})), \quad t > 0, \quad x \in (0,1), \tag{2.6}$$

$$u(0,t) = b_0(t),\ \ u(1,t) = b_1(t),\ t > 0,$$

where f and d satisfy the familiar conditions of uniform ellipticity. The second is the nonlinear Schrödinger equation

$$v_t + w_{xx} + (v^2 + w^2)w = 0,\ \ t > 0,\ \ x \in (x_L, x_R), \tag{2.7}$$

$$w_t - v_{xx} - (v^2 + w^2)v = 0,\ \ t > 0,\ \ x \in (x_L, x_R),$$

$$v_x(x,t) = w_x(x,t) = 0,\ \ x = x_L, x_R,\ t > 0.$$

Applying 3-point finite differences on nonequidistant grids ODE systems result which can be proved to satisfy (2.3), the parabolic problem in the ℓ^∞ −norm and the Schrödinger problem in the ℓ^2 −norm. □

In this paper we avoid questions concerning existence, uniqueness and smoothness of exact and numerical solutions. Hence, we suppose throughout that the two Cauchy problems at hand possess unique solutions $u(x,t)$ and $U(t)$, respectively. In addition, it is supposed that the true PDE solution is as smooth as the numerical analysis requires.

3. PRELIMINARIES ON THE FULL DISCRETIZATION

For the time integration of the ODE system (2.2) we define the implicit Runge-Kutta step $U^n \rightarrow U^{n+1}$ given by

$$U^{n+1} = U^n + \tau \sum_{i=1}^{s} b_i F(t_n + c_i\tau, Y_i),\ \ n = 0,1,..., \tag{3.1}$$

$$Y_i = U^n + \tau \sum_{j=1}^{s} a_{ij} F(t_n + c_j\tau, Y_j),\ \ i = 1(1)s,$$

where $t_0 = 0$ and U^{n+1} is the approximation to $U(t_{n+1}), t_{n+1} = t_n + \tau$. Throughout, we adopt the usual convention $c_i = a_{i1} + ... + a_{is}$, all i, and $b_1 + ... + b_s = 1$. Consequently, it is supposed that the *order of consistency p* of the integration formula of (3.1) is at least one.

Example 3.1. For future reference we already list the DIRK schemes we shall concentrate on later in the paper, viz., using Butcher's notation, *the implicit Euler rule*

$$\begin{array}{c|c} 1 & 1 \\ \hline & 1 \end{array} \qquad p = 1 \tag{3.2}$$

the implicit midpoint rule

$$\begin{array}{c|c} \frac{1}{2} & \frac{1}{2} \\ \hline & 1 \end{array} \qquad p = 2 \tag{3.3}$$

and *the 2-stage scheme*

$$\begin{array}{c|cc} \gamma & \gamma & 0 \\ 1-\gamma & 1-2\gamma & \gamma \\ \hline & \frac{1}{2} & \frac{1}{2} \end{array} \qquad \gamma = \frac{1}{2} + \frac{1}{6}\sqrt{3},\ p = 3 \tag{3.4}$$

and *the 3-stage scheme*

$$\begin{array}{c|ccc} \gamma & \gamma & 0 & 0 \\ \frac{1}{2} & \frac{1}{2}-\gamma & \gamma & 0 \\ 1-\gamma & 2\gamma & 1-4\gamma & \gamma \\ \hline & \frac{1}{24(\frac{1}{2}-\gamma)^2} & 1-\frac{1}{12(\frac{1}{2}-\gamma)^2} & \frac{1}{24(\frac{1}{2}-\gamma)^2} \end{array} \qquad \gamma = \frac{1}{2} + \frac{1}{3}\sqrt{3}\cos(\frac{\pi}{18}),\ p = 4. \tag{3.5}$$

developed independently by Nørsett [15] and Crouzeix [5]. Observe that the order of consistency p ranges from 1 to 4. Later we will show that the 2-stage and 3-stage scheme may suffer from accuracy and order reduction. □

The RK result U^{n+1} is the *full approximation* to $u_h(t_{n+1}) = r_h u(x,t_{n+1})$. Here r_h stands for the natural restriction operator on the space grid. Hence $u_h(t)$ is a vector in $\mathbb{R}^m$. We want to study the full convergence of (3.1), i.e., the behaviour of the *full discretization error*

$$\epsilon^{n+1} = u_h(t_{n+1}) - U^{n+1} \tag{3.6}$$

as both $\tau \to 0$ and $h \to 0$. Unless otherwise stated, it is supposed that τ and h are *independent parameters.* Further, for ease of presentation we restrict ourselves to constant stepsizes τ, i.e., in the limit process we take $t_N = N\tau$ fixed and suppose that $\tau \to 0$, $N \to \infty$ in such a way that $N\tau = t_N$. As ϵ is a full error it does contain the error due to discretization of the space variables. According to the MOL approach we want to treat this part separately from the error due to discretization of the time variable. For this purpose we introduce the *space truncation error*

$$\alpha(t) = F(t,u_h(t)) - \dot{u}_h(t). \tag{3.7}$$

Here $\dot{u}_h(t) = du_h(t)/dt = r_h u_t(x,t)$, i.e., the restriction of the derivative u_t of the true PDE solution u to the space grid.

Our convergence analysis is aimed at deriving full error bounds at fixed times $t_N = N\tau$ of the form

$$\|\epsilon^N\| \leqslant C_1\tau^q + C_2 \max_{0\leqslant t\leqslant t_N} \|\alpha(t)\|, \quad \forall\tau \in (0,\bar{\tau}], \quad 1 \leqslant q \leqslant p, \tag{3.8}$$

where C_1, C_2 and $\bar{\tau}$ are constants independent of τ and h. The term $C_1\tau^q$ emanates from the time integration. Clearly, the order q appearing in this bound must be smaller than or equal to p, the order of consistency of the RK formula. As C_1 and $\bar{\tau}$ are required to be independent of h (independent of the stiffness in the ODE terminology), it may very well happen that q is really smaller than p (order reduction). One can say that q is the *order uniform in h,* whereas p is the *order for fixed h.*

4. DERIVATION OF THE FULL ERROR BOUND

4.1 Convergence stability

Because our convergence analysis is centered around the semi-discrete problem, we can make fruitful use of stability results from the field of stiff ODEs [7]. Here the concept of C-stability [7,Ch.10] proves to be very useful for transferring the local errors (defined later on) to the full global error (in the definition below $\tilde{U}^n, \tilde{U}^{n+1}$ is a second numerical solution satisfying (3.1)).

Definition 4.1. Let $\|\cdot\|$ be a norm on $\mathbb{R}^m$. The integration method is called *C-stable* for the Cauchy problem (2.2) with respect to this norm, if a positive real number $\tau_0 = \tau_0(h)$ and a real constant C_0, independent of τ and h exist, such that for each $\tau \in (0,\tau_0]$ and each $U^n, \tilde{U}^n \in \mathbb{R}^m$

$$\|\tilde{U}^{n+1} - U^{n+1}\| \leqslant (1+C_0\tau)\|\tilde{U}^n - U^n\|. \quad \square \tag{4.1}$$

C-stability is an abbreviation for *convergence stability* and is linked with stability in the Lax-Richtmeyer sense [16] and, more closely with stability in the sense of Kreiss [13] (sometimes referred to as strong stability [16]). If $C_0 \leqslant 0$ and we think of U^n, as being a numerical solution, and of $\tilde{U}^n$ as being a perturbation of U^n, then (3.10) shows that the perturbation will not increase in time. The bound (3.10) then provides the definition of contractivity, also called *computing stability*, a concept which plays a major role in recent developments in ODES [7]. If $C_0 > 0$, we allow an increase in the difference $\tilde{U}^n - U^n$. In this case C-stability is mainly useful in the convergence analysis and not as a concept of computing stability. Notice that C-stability is a property for *nonlinear* problems. In general τ_0 may decrease with h. However, for the given DIRK schemes we have a fixed bound $\tau_0(h)$ for τ, under the hypothesis (2.3):

Theorem 4.1. Let hypothesis (2.3) be true for a given norm $\|\cdot\|$ on $\mathbb{R}^m$. Then (i) The implicit Euler method is C-stable for this norm (ii) The implicit midpoint rule and the 2-stage and 3-stage schemes (3.4) and (3.5) are C-stable if $\|\cdot\|$ is an inner product norm. Further, for all four schemes τ_0 and C_0 depend solely on $\mu_{\max}$. $\square$

The proof of this theorem can be found in the literature on nonlinear stiff ODEs (see the survey [7], §2.4 for (i), §7.4 for (ii)). The result for implicit Euler goes back to Desoer & Haneda [8]. The C-stability of implicit midpoint has been proved by various authors and is in fact known for a long time. The result for the 2-stage and 3-stage scheme is of a more recent date and can be concluded from the general Th.7.4.2 in [7]: *if an algebraically stable RK method is BSI-stable, then it is C-stable.* At this place we should like to mention that the proof of BSI-stability of the 2-stage and 3-stage DIRK scheme, given in [7], is largely due to Montijano [13].

It shall be clear now that for the DIRK schemes applied to the problem classes (2.1)-(2.2) satisfying (2.3), stability in the sense of Definition 4.1 is guaranteed. We now leave the subject of C-stability and shall proceed with the examination of a recurrence for the full error ϵ where, in the usual way, C-stability takes care of transferring full local errors to ϵ.

4.2. A recurrence for the full error

We consider the Runge-Kutta step $U^n \to U^{n+1}$ given by (3.1) and the perturbed fictitious step $u_h(t_n) \to u_h(t_{n+1})$ given by

$$u_h(t_{n+1}) = u_h(t_n) + \tau \sum_{i=1}^{s} b_i F(t_n + c_i\tau, u_h(t_n + c_i\tau)) + r_0, \tag{4.2}$$

$$u_h(t_n + c_i\tau) = u_h(t_n) + \tau \sum_{j=1}^{s} a_{ij} F(t_n + c_j\tau, u_h(t_n + c_j\tau)) + r_i, \quad i = 1(1)s.$$

The (specific) perturbations r_i are residuals depending exclusively on the true PDE solution u_h and on the space truncation error α. For, using (3.7),

$$r_0 = u_h(t_{n+1}) - u_h(t_n) - \tau \sum_{i=1}^{s} b_i \dot{u}_h(t_n + c_i\tau) - \tau \sum_{i=1}^{s} b_i \alpha(t_n + c_i\tau), \tag{4.3}$$

$$r_i = u_h(t_n + c_i\tau) - u_h(t_n) - \tau \sum_{j=1}^{s} a_{ij} \dot{u}_h(t_n + c_j\tau) - \tau \sum_{j=1}^{s} a_{ij} \alpha(t_n + c_j\tau), \quad i = 1(1)s.$$

By straightforward Taylor expansion of u_h it follows that integers $p_i \geqslant 1$ (recall the convention made for (3.1)) and positive reals d_i, $i = 0(1)s$, exist such that uniformly in n

$$\|r_0\| \leqslant d_0 \tau^{p_0+1} + \tau \sum_{i=1}^{s} |b_i| \, \|\alpha(t_n + c_i\tau)\|, \tag{4.4}$$

$$\|r_i\| \leqslant d_i \tau^{p_i+1} + \tau \sum_{j=1}^{s} |a_{ij}| \, \|\alpha(t_n + c_j\tau)\|, \quad i = 1(1)s.$$

We note that all d_i are determined exclusively by bounds for one or more of the derivatives $\ddot{u}_h$, $\dddot{u}_h$,.... In the work of Frank, Schneid & Ueberhuber [9,10], the minimum of p_i, $\tilde{p}$ say, is called the *stage order*.

Let us return to formulas (4.2) and subtract (3.1). If we define the intermediate errors $\epsilon_i = u_h(t_n + c_i\tau) - Y_i$, we then get the error scheme

$$\epsilon^{n+1} = \epsilon^n + \tau \sum_{i=1}^{s} b_i A_i \epsilon_i + r_0, \tag{4.5a}$$

$$\epsilon_i = \epsilon^n + \tau \sum_{j=1}^{s} a_{ij} A_j \epsilon_j + r_i, \quad i = 1(1)s, \tag{4.5b}$$

where, according to the mean value theorem for vector functions,

$$A_i = \int_0^1 F'(t_n + c_i\tau, \theta u_h(t_n + c_i\tau) + (1-\theta) Y_i) d\theta, \quad i = 1(1)s. \tag{4.6}$$

For convenience we suppress the dependence of A_i on n, like we did for Y_i, r_i and ϵ_i. Supposing that (4.5b) can be solved for $\epsilon_1, \ldots, \epsilon_s$ we thus arrive at the full error recurrence which is of the familiar form

$$\epsilon^{n+1} = R^{(n)}\epsilon^{n} + \beta^{n+1}, \tag{4.7}$$

with $R^{(n)}$ as the *amplification matrix* and β^{n+1} as the *full local error.*

The solution of the algebraic system (4.5b) is rather complicated for the general method (3.1) (see [7], Ch.5 for an extensive discussion), but fairly simple for DIRK schemes since then $a_{ij} = 0$, $j > i$, $i = 1(1)s$. We only need to assess the invertability of the matrices $I - \tau a_{ii}A_i$.

Lemma 4.1. Suppose (2.3) for a given norm $\|\cdot\|$. Then $I - \gamma\tau A_i$ is invertible for all $\tau > 0$ satisfying $\gamma\tau\mu_{\max} < 1$ while

$$\|(I-\gamma\tau A_i)^{-1}\| \leqslant \frac{1}{1-\gamma\tau\mu_{\max}}, \quad \|\gamma\tau A_i(I-\gamma\tau A_i)^{-1}\| \leqslant 1 + \frac{1}{1-\gamma\tau\mu_{\max}}. \tag{4.8}$$

Proof. The proof follows from known properties of the logarithmic norm (Dahlquist [6]). Also given in [7], Lemma 1.5.4 and Theorem 2.4.1. □

It follows that for the integration schemes and problem class under consideration the recurrence (4.7) is well defined. We may also conclude that for the DIRK schemes (3.2)-(3.5) the C-stability inequality

$$\|\epsilon^{n+1}\| \leqslant (1+C_0\tau)\|\epsilon^{n}\| + \|\beta^{n+1}\|, \quad \forall\tau \in (0,\tau_0], \tag{4.9}$$

holds due to Th.4.1 (provided the correct norm is chosen). This statement can be understood from the observation that if we subtract (3.1) from the perturbed RK step $\tilde{U}^{n} \to \tilde{U}^{n+1}$, where we only consider equal perturbations like in Def.4.1, that then $\tilde{U}^{n+1} - U^{n+1} = R^{(n)}(\tilde{U}^{n} - U^{n})$ provided the definition of A is appropriately changed. Consequently, as C_0 is independent of τ and h, for finding error bounds of type (3.8) it suffices to prove that for the local error β^{n+1} a similar bound exist with the right hand side *multiplied* by τ.

By using (4.4) and (4.8) such local error bounds can be obtained in a straightforward manner for any DIRK scheme from the explicitly available expression for β^{n+1}. Rather than considering the general DIRK scheme, we shall carry out the computation for each of the four schemes (3.2)-(3.5). This enables us to discuss in greater detail the emerging order reduction phenomena. Finally we want to emphasize that the ideas behind the presented error analysis are borrowed from the B-convergence theory for stiff ODEs due to Frank, Schneid & Ueberhuber [9,10]. However, the derivation presented here is a bit shorter than in [9,10] and, in our opinion, also slightly more transparant. More details concerning this point can be found in a forthcoming paper with K. Burrage and W. Hundsdorfer [3].

4.3. The first order implicit Euler scheme (3.2)

From (4.5) we immediately can write down the error recurrence (4.7), i.e.

$$\epsilon^{n+1} = (I-\tau A_1)^{-1}\epsilon^n + \beta^{n+1}, \tag{4.10}$$

$$\beta^{n+1} = (I-\tau A_1)^{-1} r_0, \quad r_0 = u_h(t_{n+1}) - u_h(t_n) - \tau \dot{u}_h(t_{n+1}) - \tau \alpha(t_{n+1}).$$

Hence, according to (4.8), for u_h in C^2,

$$\|\beta^{n+1}\| \leqslant \frac{1}{1-\tau\mu_{\max}}\left(\frac{1}{2}M_2\tau^2 + \tau\|\alpha(t_{n+1})\|\right), \quad \tau\mu_{\max} < 1, \tag{4.11}$$

where M_2 is an upper bound for $\|\ddot{u}_h(t)\|$. In view of the C-stability of implicit Euler, the full error bound (3.8) exists with $q = p = 1$ (no order reduction). It shows convergence of order one in time as $\tau, h \to 0$ in any way and for any norm for which (2.3) holds. An interesting feature is that only $\ddot{u}_h$ enters into the bound. We emphasize that this convergence result for implicit Euler is well known in the PDE and stiff ODE literature.

4.4. The second order implicit midpoint scheme (3.3)

The error scheme (4.5) now reads

$$\epsilon^{n+1} = \epsilon^n + \tau A_1 \epsilon_1 + r_0, \; \epsilon_1 = \epsilon^n + \frac{1}{2}\tau A_1 \epsilon_1 + r_1, \tag{4.12}$$

and the local error β^{n+1} is given by

$$\beta^{n+1} = (I - \frac{1}{2}\tau A_1)^{-1}\tau A_1 r_1 + r_0, \tag{4.13}$$

$$r_0 = u_h(t_n + \tau) - u_h(t_n) - \tau\dot{u}_h(t_n + \frac{1}{2}\tau) - \tau\alpha(t_n + \frac{1}{2}\tau),$$

$$r_1 = u_h(t_n + \frac{1}{2}\tau) - u_h(t_n) - \frac{1}{2}\tau\dot{u}_h(t_n + \frac{1}{2}\tau) - \frac{1}{2}\tau\alpha(t_n + \frac{1}{2}\tau).$$

Using the second of inequalities (4.8), and (4.4), we find for u_h in C^3,

$$\|\beta^{n+1}\| \leqslant cd_1\tau^2 + d_0\tau^3 + \tau(1+\frac{1}{2}c)\|\alpha(t_n + \frac{1}{2}\tau)\|, \quad \frac{1}{2}\tau\mu_{\max} < 1,$$

where $c = 2(1+1/(1-\frac{1}{2}\tau\mu_{\max}))$. Consequently, for inner product norms the existence of a full error bound (3.8) has been shown, but only with q equal to the stage order $\tilde{p} = 1$. This result suggests that implicit midpoint may suffer from order reduction, unless the differential equation meets an additional requirement (cf. class D2 in [10]; in our setting this condition reads

$$\|F(t, u_h + \delta\tau^2\ddot{u}_h) - F(t, u_h)\| = O(\tau^2), \; \delta = -\frac{1}{8}, \tag{4.14}$$

uniformly in h). Fortunately, this suggestion is false. The situation is that the local error β^{n+1} indeed may suffer from a reduction ([9,10],[7],Ch.7), but, quite unexpectedly, the global error ϵ^{n+1} does not. This last point has been proved by Stetter [16] and just recently by Kraaijevanger [11] (see also Axelsson [1]). Kraaijevanger's proof fits best in our setting. His idea is to treat an appropriately perturbed error scheme where the defect of the intermediate stage has been removed.

Write r_0^n, r_1^n for r_0, r_1. Let $\tilde{\epsilon}^n = \epsilon^n + r_1^n$. Then $\tilde{\epsilon}^n$ satisfies

$$\tilde{\epsilon}^{n+1} = \tilde{\epsilon}^n + \tau A_1 \epsilon_1 + \tilde{\beta}^{n+1}, \quad \epsilon_1 = \tilde{\epsilon}^n + \frac{1}{2}\tau A_1 \epsilon_1 \tag{4.15}$$

and $\tilde{\beta}^{n+1} = r_1^{n+1} - r_1^n + r_0^n$ can be interpreted as a (perturbed) truncation error. Because $\tilde{\epsilon}^{n+1} = R^{(n)}\tilde{\epsilon}^n + \tilde{\beta}^{n+1}$, these new local errors, say for $n = 0(1)N-1$, can be transferred to $\tilde{\epsilon}^N$ in the standard way using the C-stability property. Herewith r_1^N should be defined as the zero vector so that $\epsilon^N = \tilde{\epsilon}^N$ and $\tilde{\beta}^N = r_0^{N-1} - r_1^{N-1}$. Note that $\tilde{\epsilon}^0 = r_1^0$ if $\epsilon^0 = \underline{0}$. Neglecting α in r_0^n, r_1^n it is easily verified that $\|\tilde{\epsilon}^0\|, \|\tilde{\beta}^N\| \leq M_2\tau^2/8$ and, for $n = 0(1)N-2$, $\|\tilde{\beta}^{n+1}\| \leq M_3\tau^3/12$. Here M_2, M_3 represent bounds for $\|\ddot{u}_h\|$ and $\|\dddot{u}_h\|$, respectively. In this way the *global* error bound (3.8) is proved with $q = p = 2$ (no reduction). Noteworthy is that C_1 is determined exclusively by M_2 and M_3. For more details, a.o. concerning variable stepsizes τ and the trapezoidal rule, we refer to [11].

4.5. <u>The third and fourth order DIRK schemes (3.4),(3.5)</u>

In view of the experiences in the field of stiff ODEs, see e.g. [7], §7.5 for numerical experiments with (3.5), we must reckon with eventual order reduction when a DIRK scheme of higher order is used for the time integration of a PDE. We shall discuss this now for the 3-rd and 4-th order schemes (3.4) and (3.5). In our analysis we hereby concentrate on (3.4) and remark that the 4-th order scheme can be dealt with in the same manner. For (3.4) the error scheme (4.5) reads

$$\epsilon^{n+1} = \epsilon^n + \frac{1}{2}\tau A_1\epsilon_1 + \frac{1}{2}\tau A_2\epsilon_2 + r_0, \tag{4.16}$$

$$\epsilon_1 = \epsilon^n + \gamma\tau A_1\epsilon_1 + r_1,$$

$$\epsilon_2 = \epsilon^n + (1-2\gamma)\tau A_1\epsilon_1 + \gamma\tau A_2\epsilon_2 + r_2,$$

and the local error β^{n+1} is given by

$$\beta^{n+1} = r_0 + (\frac{1}{2}B_1 + \frac{1}{2}(1-2\gamma)B_2B_1)r_1 + \frac{1}{2}B_2r_2. \tag{4.17}$$

For convenience of notation we introduced the abbreviation $B_i = (I - \gamma\tau A_i)^{-1}\tau A_i$. The residuals r_i (cf.(4.3)) satisfy, for u_h in C^4 and for any γ,

$$r_0 = (-\frac{1}{12} + \frac{1}{2}\gamma - \frac{1}{2}\gamma^2)\tau^3\dddot{u}_h(t_n) + O(\tau^4) - \frac{1}{2}\tau\alpha(t_n + \gamma\tau) - \frac{1}{2}\tau\alpha(t_n + (1-\gamma)\tau), \tag{4.18}$$

$$r_1 = \frac{1}{2}\gamma^2\tau^2\ddot{u}_h(t_n) + \frac{1}{3}\gamma^3\tau^3\dddot{u}_h(t_n) + O(\tau^4) - \gamma\tau\alpha(t_n + \gamma\tau),$$

$$r_2 = (-\frac{1}{2} + 3\gamma - \frac{7}{2}\gamma^2)\tau^2\ddot{u}_h(t_n) - (\frac{1}{6} - \gamma + \gamma^2 + \frac{1}{3}\gamma^3)\tau^3\dddot{u}_h(t_n) +$$

$$O(\tau^4)-(1-2\gamma)\tau\alpha(t_n+\gamma\tau)-\gamma\tau\alpha(t_n+(1-\gamma)\tau).$$

If $\gamma = \frac{1}{2}\pm\frac{1}{6}\sqrt{3}$, the τ^3-term of r_0 vanishes. This value of γ corresponds to the order $p=3$. Using the stability argument (the scheme is C-stable for $\gamma \geqslant 1/4$) and the boundedness of B_i it thus follows that for the DIRK scheme (3.4) a global error bound (3.8) exists with $q \geqslant 1$, i.e., $q \geqslant \tilde{p}$, the stage order. This result is disappointing as $p=2$ for any γ and $p=3$ for $\gamma = \frac{1}{2} \pm \frac{1}{6}\sqrt{3}$. For problems satisfying the condition (4.14), the order $q=2$ is obtained. However, the constant C_1 in (3.8) then will depend also on the size of $F(t,u_h+\delta\ddot{u}_h)-F(t,u_h)$ and no longer exclusively on the smoothness of u_h.

Extensive numerical experiments has led us to the *conjecture* that ϵ *always* satisfies a bound (3.8) with $q \geqslant 2$, rather than $q \geqslant 1$, although β may show a reduction which is more in line with $q \geqslant 1$. This means that we are in a similar situation as with the implicit midpoint rule. Note, however, that in the present case ϵ does suffer from a reduction. In fact, experiments showing a virtual 2-nd order in time for the global error are easily conducted.

When attempting to prove the *conjecture* the first approach which comes to mind is that of analysing an appropriate perturbation of (4.16), like Kraaijevanger did for the midpoint rule. A little reflection shows that this is easily done if the leading terms of r_1 and r_2 are equal, which is the case only if $\gamma = 1/2, 1/4$. For other values of γ the perturbation approach seems to lead to a rather complicated analysis, but is feasible for problems of the semi-linear type $\dot{U}=AU+G(t,U)$ [3].

A case study. We shall now outline an alternative method of proof for our conjecture for the constant coefficient problem

$$\dot{U} = AU + G(t). \qquad (4.19)$$

The method of proof can be extended to problems of the above semi-linear type $\dot{U} = AU + G(t,U)$ where $\|G'(t,\zeta)\| < \infty$ uniformly in h.

Consider the error scheme (4.5). Let us write r_i^n for r_i. Put $\tilde{\epsilon}^n = \epsilon^n + r_1^n$. Then (note that $R^{(n)}, B_1, B_2$ are independent of n in this case)

$$\tilde{\epsilon}^N = R^N\tilde{\epsilon}^0 + \sum_{n=0}^{N-1} R^n\tilde{\beta}^{n+1}, \quad \tilde{\beta}^{n+1} = (r_0^n - r_1^n + r_1^{n+1}) + \frac{1}{2}B(r_2^n - r_1^n), \qquad (4.20)$$

which we write as

$$\tilde{\epsilon}^N = R^N\tilde{\epsilon}^0 + \sum_{n=0}^{N-1} \frac{1}{2}R^nB\hat{r}^n + \sum_{n=0}^{N-1} R^n(\tilde{\beta}^{n+1} - \frac{1}{2}B\hat{r}^n), \qquad (4.21)$$

where $\hat{r}^n$ is the difference of the leading terms of r_1 and r_2, i.e., $\hat{r}^n = (\frac{1}{2}-3\gamma+4\gamma^2)\tau^2\ddot{u}_h(t_n)$. Using the stability argument on R and the boundedness of B it thus can be seen that for proving (3.8) with $q=2$ it suffices to prove that the second term, say S, satisfies $\|S\| \leqslant C\tau^2$ for all $\tau(0,\bar{\tau}]$ uniformly in h.

In what follows we now consider the most simple case where $\ddot{u}_h$ is constant, i.e., $\hat{r}^n = \hat{r}^0$ for all n. Also suppose that $I - R$ is regular (both restrictions are not essential and can be removed). Then S can be brought in the form

$$S = (I - R^N)(I - R)^{-1}\frac{1}{2}B\hat{r}^0 \tag{4.22}$$

$$= \frac{1}{2}(\frac{1}{2} - 3\gamma + 4\gamma^2)(I - R^N)(I + (\frac{1}{2} - 2\gamma)\tau A)^{-1}(I - \tau A)\tau^2\ddot{u}_h(0),$$

where we used the expression $R = I + B + \frac{1}{2}(1-2\gamma)B^2$. Again using the stability argument to cope with R^N and inequalities (4.8) for the rational expression in τA, finally shows that S is of second order in τ, uniformly in h, for all $\gamma > 1/4$ (for $\gamma = 1/4, 1/2$ we have $S = 0$).

For clarity, the essence of the proof is to bound the whole series S rather than its individual terms $R^n B\hat{r}^n$. The philosophy here it to attack directly the global error rather than following the standard approach of the convergence analysis which consists of first bounding locally and then adding all bounds via the stability argument. We also emphasize that no additional condition, such as (4.14), has been made and that the constant C_1 in the resulting bound (3.8) for ϵ^N is determined exclusively by μ_{max} and bounds for $d^2u_h/dt^2, d^3u_h/dt^3$ and d^4u_h/dt^4 (only if $\gamma = \frac{1}{2} + \frac{1}{6}\sqrt{3}$). Note that for problem (4.19), (4.14) implies that A and u_h should satisfy $A\ddot{u}_h = 0(1)$ uniformly in h. In the example below we will show, that already for simple PDE problems, leading to (4.19), (4.14) is a too severe restriction. □

Example 4.1. The objective of this example is two-fold. We want to show, for a concrete but simple problem, that the local error β indeed may suffer from more reduction than the global error, thus motivating the global approach we followed in the case study. In the second place we want to illustrate in which cases order reduction is to be expected for the method (3.4) (and (3.5)).

Let the semi-discrete system be of type (4.19) and suppose that u_h is a quadratic polynomial (this restriction is not essential and can be removed). Let $\gamma = \frac{1}{2} + \frac{1}{6}\sqrt{3}$, so we have $p = 3$ in (3.4). The local error $\hat{\beta}^{n+1}$ given by (4.17) then takes the form $\beta^{n+1} = \hat{\beta} +$ space error, where $\hat{\beta}$ is the time error

$$\hat{\beta} = \frac{1}{4}\gamma^2(1-2\gamma)B^2\tau^2\ddot{u}_h(0) = \frac{1}{4}\gamma^2(1-2\gamma)(I - \gamma\tau A)^{-2}\tau^4 A^2\ddot{u}_h(0), \tag{4.23}$$

which is independent of n. We now confine our attention to $\hat{\beta}$. Clearly, in order that $\hat{\beta} = O(\tau^{p+1}) = O(\tau^4)$, uniformly in h, it is sufficient and necessary that $(I - \gamma\tau A)^{-2}A^2\ddot{u}_h(0) = 0(1)$, uniformly in τ and h. However, this boundedness condition is rather restrictive and essentially requires that $A^2\ddot{u}_h(0) = O(1)$, uniformly in h. As A contains negative powers of h, u_{tt} then should not only be smooth enough in x, but also satisfy the boundary conditions imposed by A^2. However these b.c. are not natural (see also [2],p.7). To show this we consider the simple parabolic equation

$$u_t = u_{xx} + g(x,t), \quad t > 0, \quad 0 \leqslant x \leqslant 1, \tag{4.24}$$

with the exact solutions (imposed by adapting $g(x,t)$)

$$u(x,t) = t^2x(1-x) \text{ and (homogeneous) Dirichlet b.c.,} \tag{4.25a}$$

$$u(x,t) = t^2(x+\frac{1}{2})(\frac{3}{2}-x) \text{ and (inhomogeneous) Dirichlet b.c..} \tag{4.25b}$$

For the discretization in space we select 2-nd order finite differences on a uniform grid. Then (4.24) is converted into (4.19) where A is the finite difference matrix

$$A = \frac{1}{h^2}\begin{bmatrix} -2 & 1 & & & \\ 1 & -2 & 1 & & \\ & \ddots & \ddots & \ddots & \\ & & 1 & -2 & 1 \\ & & & 1 & -2 \end{bmatrix}_{m\times m}, \quad h = \frac{1}{m+1}. \tag{4.26}$$

The definition of G in (4.19) shall be clear. Note that the discretization in space is exact since u is a quadratic polynomial in x, in both cases. Hence, $\beta^{n+1} = \hat{\beta}$, $n = 0, 1, \ldots$

Now let $\tau = h \to 0$. Then the following asymptotic behaviour is observed:

$$\|\hat{\beta}\|_2 \sim \begin{matrix} \tau^{3.25} \text{ for } (4.25a), \\ \tau^{2.25} \text{ for } (4.25b), \end{matrix} \tag{4.27}$$

where $\|\cdot\|_2 = (h<.,.>)^{1/2}$, the standard l^2 norm. In the homogeneous case we have a reduction in local order from 4 to 3.25, and in the inhomogeneous case even from 4 to 2.25.

In the *homogeneous case* the reduction originates from the fact that u_{ttxx} is not zero on the boundary $x = 0,1$. To see this, u_{ttxx} is approximated by $A\ddot{u}_h$. Here, $A\ddot{u}_h(0) = 2[-2, \ldots, -2]^T$. However, this implies that $A^2\ddot{u}_h(0) = 2[2h^{-2}, 0, \ldots, 0, 2h^{-2}]^T$, i.e., the nearby boundary components of $A^2\ddot{u}_h$ are unbounded. Fortunately, these extremely large boundary errors are smeared out and diminished through the multiplication by $(I-\gamma\tau A)^{-2}$. In passing we note that $\hat{\beta} = O(\tau^3)$, uniformly in h, as $A\ddot{u}_h(0) = 0(1)$ (see (4.8)).

In the *inhomogeneous case* we have a similar situation, but here the reduction is larger because already u_{tt} does not vanish at $x = 0,1$ which implies that the nearby boundary components of $A\ddot{u}_h$, and $A^2\ddot{u}_h$, are unbounded in h. Notice that now condition (4.4) does not hold and that $\hat{\beta} = O(\tau^2)$, uniformly in h, as $\ddot{u}_h(0) = 0(1)$.

At first sight one might think now that we have to face a reduction in global order from , respectively, 3 to 2.25 and 3 to 1.25 as $\tau = h \to 0$. However, a direct consequence from our case study is that for both solutions (4.25) the global error is at least $O(\tau^2)$, uniformly in h. To illustrate this numerically we have integrated the problems (4.24)-(4.26) in time over the interval [0,1] using the 3-rd and 4-th order method (the latter was applied for the sake of comparison). Table 4.1 shows the quantity

$$p_2(N) = \log_2\|\epsilon^N\|_2 \,/\, \|\epsilon^{2N}\|_2, \quad N\tau = 1, \tag{4.28}$$

i.e., the *order of accuracy* measured using $\tau = h = N^{-1}, (2N)^{-1}$. Recall that no space error is present. The floating point numbers are $\|\epsilon^{10}\|_2$.

Table 4.1. Order test for methods (3.4), (3.5) applied to problems (4.24)-(4.26). The left table corresponds to the homogeneous b.c., the right table to the inhomogeneous ones.

N	10	20	40	80	160
(3.4)	1.6_{10}^{-4}	2.56	2.72	2.83	2.90
(3.5)	8.7_{10}^{-4}	2.99	3.28	3.40	3.33

N	10	20	40	80	160
(3.4)	8.2_{10}^{-4}	2.34	2.34	2.29	2.26
(3.5)	5.0_{10}^{-4}	2.38	2.25	2.21	2.22

We see that in the case of the homogeneous b.c., p_2 tends to $p=3$ for method (3.4) (no virtual reduction visible), whereas for method (3.5) the p_2-values indicate clearly that reduction occurs. In the case of the inhomogeneous b.c. both methods suffer from reduction. Noticeable is that it is larger for the 4-th order method (3.5) (from 4 to approximately 2.2). This experiment shows that even for simple parabolic problems with smooth solutions and inhomogeneous b.c. there may be no advantage at all in using high order in time. Finally, it is worthwhile to remark that the same results are found when we keep h fixed and consider a finite, realistic range of τ-values. □

Remark 4.1. Brenner, Crouzeix & Thomée [2] reported earlier on the phenomenon of order reduction for RK methods applied to PDEs. They restrict their analysis to constant coefficient linear problems (in Banach space) and examine only reduction of the local error. They also use problem (4.24)-(4.25) as an example. □

Remark 4.2. The case study and the example treated in this paragraph were meant to give insight into the local and global error behaviour of higher order DIRK schemes. It is noted that a proof of our conjecture that $q \geqslant 2$ in (3.8) has not yet been obtained for the general nonlinear problem (2.1)-(2.3). The method of proof followed in the case study can probably not extended to this general nonlinear problem (see also [3]). In the example we have shown the origin of the order reduction. We want to remark that the restriction to constant A is not essential. Also for A time dependent, thus covering the most general situation, an expression similar to (4.23) can be derived from (4.17). However, this expression is lengthy and complicated and renders no more insight. □

Example 4.2. The objective of this example is to call attention for another source of inaccuracy, viz., *non-smooth coefficients* in the PDE operator (non-smooth in the sense of having large gradients). It is best illustrated from a concrete, simple problem. Consider the parabolic equation

$$u_t = (d(x)u_x)_x + g(x,t), \quad t > 0, \quad 0 \leqslant x \leqslant 1, \tag{4.29}$$

with Dirichlet boundary conditions. Let u be a quadratic polynomial in t. Any (finite difference) semi-discrete approximation takes the form (4.19) and, like in Example 4.1, the time error part $\hat{\beta}$ of the local error (4.17) is given by (4.23).

Now examine the grid functions $A\ddot{u}_h$, $A^2\ddot{u}_h$. Clearly, $A\ddot{u}_h$ represents an approximation to $(du_{ttx})_x$ and

$A^2\ddot{u}_h$ to $(d(du_{ttx})_{xx})_x$. Next suppose that d has much larger gradients than u so that

$$|(d(du_{ttx})_{xx})_x| >> |(du_{ttx})_x| >> |u_{tt}|,$$

which will imply that for all components $(\cdot)_j$, and for any realistic value of h,

$$|(A^2\ddot{u}_h)_j| >> |(A\ddot{u}_h)_j| >> |(\ddot{u}_h)_j|.$$

This observation suggests that the bound $\|\hat{\beta}\| \leq C\tau^2$, derived from the expression

$$\hat{\beta} = \tfrac{1}{4}\gamma^2(1-2\gamma)(\tau^2A^2(I-\gamma\tau A)^{-2})\tau^2\ddot{u}_h(0), \tag{4.23'}$$

and thus with C *independent of the non-smooth coefficient* d, will be in better accord with the true error behaviour than a higher order bound where the constant involved does depend on d.

To test this we have repeated the numerical experiment of Example 4.1 using the *non-smooth coefficient* $d(x) = (2+x)^8$ and the solutions (4.25). For the discrétization in space we here used the standard 4-th order finite difference formula, except at the nearby boundary points where a 3-rd order approximation was applied. Table 4.2 shows the results in exactly the same way as Table 4.1. These results indeed reveal a distinct *2-nd order behaviour* for both solutions (4.25) (and both methods). □

Table 4.2. (same information as in table 4.1).

N	10	20	40	80	160	N	10	20	40	80	160
(3.4)	$2.1_{10}-4$	1.62	1.90	1.98	2.00	(3.4)	$1.3_{10}-3$	1.86	1.97	1.99	2.00
(3.5)	$1.7_{10}-4$	1.60	1.88	1.97	2.00	(3.5)	$1.1_{10}-3$	1.90	1.97	1.99	2.00

5. A NUMERICAL STUDY

Our DIRK schemes of order $p > 2$ do suffer from order reduction as the numerical experiments of §4 clearly illustrate. One then should question whether the extra computational work needed to reach this order p pays off. The answer to this question is not easy to give since in general there are many factors involved (type of problem, level of accuracy, stability, eventual stepsize control , iteration strategy). Despite this inherent uncertainty we have conducted numerical experiments on some more problems in an attempt to supplement our theoretical results with a conclusion which is of some value to the numerical practice. The present section is devoted to three of these problems (scalar parabolic PDEs from practice, but with smooth solutions). For the sake of comparison all four DIRK methods discussed in this paper were applied.

For the discretization in space we used a uniform grid and the standard 4-th order finite difference technique, except at the nearby boundary points where a 3-rd order formule was implemented. Further in all cases $h = \tau_B$, so h decreases with the stepsize τ. In the tables of result we have listed the full error

$\|\epsilon^N\|_2, N\tau = T$ and the quantity $p_2(N)$ given by (4.28). In each experiment we selected one basic stepsize τ_B and then used $\tau = \tau_B$ for the two 1-stage schemes and $\tau = 2\tau_B, 3\tau_B$ for the 2-stage and 3-stage scheme, respectively, thus accounting the extra work of the latter ones.

Noteworthy is that according to this way of presentation, the 2-stage (DIRK2) and 3-stage (DIRK3) method are considered to be 2 and 3 times as expensive as EULER and MIDPOINT, respectively. Thus we tacitly assume that the costs per stage are equal, for all four methods and all stepsizes. For nonlinear problems this may be somewhat in favour of the higher stage methods because these become attractive only when they are capable of yielding sufficient accuracy for relatively large stepsizes. In order to reach this accuracy it then may be necessary, in practice, to do some more Newton iterations per stage, which, to some extent, then annihilates the advantage of a greater stepsize.

Problem I. The Burger's equation

$$u_t = \nu u_{xx} - u u_x, \quad 0 < t \leqslant T = 1, \quad 0 < x < 1. \tag{5.1}$$

This equation has been studied by many authors (e.g. by Varah [17]). For $\nu << 1$, steep gradients may exist in the solution u. We used the "large" value $\nu = 0.1$ and defined initial and Dirichlet boundary values from the exact solution given by Whitham [19], Ch.4.

$$u(x,t) = 1 - 0.9 \frac{r_1}{r_1 + r_2 + r_3} - 0.5 \frac{r_2}{r_1 + r_2 + r_3}, \tag{5.2}$$

where $r_1 = e^{-\frac{x-.5}{20\nu} - \frac{99t}{400\nu}}$, $r_2 = e^{-\frac{x-.5}{4\nu} - \frac{3t}{16\nu}}$, $r_3 = e^{-\frac{x-3/8}{2\nu}}$.

Table 5.1. Results for Burger's equation (5.1)-(5.2).

τ_B	$EULER(\tau_B)$		$MIDPOINT(\tau_B)$		$DIRK2(2\tau_B)$		$DIRK3(3\tau_B)$	
	$\|\epsilon^N\|_2$	p_2	$\|\epsilon^N\|_2$	p_2	$\|\epsilon^N\|_2$	p_2	$\|\epsilon^N\|_2$	p_2
1/24	1.2_{10}^{-3}		4.6_{10}^{-5}		5.6_{10}^{-5}		8.8_{10}^{-5}	
1/48	6.0_{10}^{-4}	1.0	1.2_{10}^{-5}	2.0	9.8_{10}^{-6}	2.52	1.4_{10}^{-5}	2.63
1/96	3.0_{10}^{-4}	1.0	2.9_{10}^{-6}	2.0	1.8_{10}^{-6}	2.43	2.8_{10}^{-6}	2.35
1/192	1.5_{10}^{-4}	1.0	7.3_{10}^{-7}	2.0	3.6_{10}^{-7}	2.34	5.9_{10}^{-7}	2.25
1/384	7.6_{10}^{-4}	1.0	1.8_{10}^{-7}	2.0	7.4_{10}^{-8}	2.28	1.3_{10}^{-7}	2.23

The results, collected in Table 5.1, reveal a distinct order reduction of the 3-rd order DIRK2 and the 4-th order DIRK3. In contrast, the order one and two of EULER and MIDPOINT clearly shows up. An interesting observation is that the p_2-values of the 3-rd and 4-th order method again are nearly equal (compare with Table 4.1, right table, and Table 4.2) and approach 2. A consequence is that these two methods do not perform better than MIDPOINT.

Problem II. Again the Burger's equation

$$u_t = \pi^{-2}\nu u_{xx} - \pi^{-1}uu_x,\ \ 0 < t \leqslant T = 1,\ \ 0 \leqslant x \leqslant 1, \tag{5.3}$$

but now with homogeneous boundary conditions $u(0,t) = u(1,t) = 0$ and with the initial function $u(x,0) = u_0\sin(\pi x)$. The exact solution of this problem was obtained by Cole [4] and reads

$$u(x,t) = \frac{4\nu \sum_{n=1}^{\infty} e^{-\nu n^2 t} \mathcal{I}_n(\frac{u_0}{2\nu})\sin(n\pi x)}{\mathcal{I}_0(\frac{u_0}{2\nu}) + 2\sum_{n=1}^{\infty} e^{-\nu n^2 t}\mathcal{I}_n(\frac{u_0}{2\nu})\cos(n\pi x)}, \tag{5.4}$$

where $\mathcal{I}_n(y)$ is the modified Bessel function of the first kind. In this example we chose $\nu = \pi^2/10$ and $u_0 = 1$. Results are given in Table 5.2

Table 5.2. Results for Burger's equation (5.3)-(5.4).

τ_B	$EULER(\tau_B)$		$MIDPOINT(\tau_B)$		$DIRK2(2\tau_B)$		$DIRK3(3\tau_B)$	
	$\|\epsilon^N\|_2$	p_2	$\|\epsilon^N\|_2$	p_2	$\|\epsilon^N\|_2$	p_2	$\|\epsilon^N\|_2$	p_2
1/24	8.1_{10}^{-3}		6.5_{10}^{-5}		3.0_{10}^{-5}		6.4_{10}^{-5}	
1/48	4.1_{10}^{-3}	.97	1.6_{10}^{-5}	2.06	4.8_{10}^{-6}	2.67	7.8_{10}^{-6}	3.04
1/96	2.1_{10}^{-3}	.99	3.9_{10}^{-6}	2.01	6.9_{10}^{-7}	2.79	7.3_{10}^{-7}	3.42
1/192	1.0_{10}^{-3}	.99	9.7_{10}^{-7}	2.00	9.2_{10}^{-8}	2.91	5.7_{10}^{-8}	3.68
1/384	5.2_{10}^{-4}	1.00	2.4_{10}^{-7}	2.00	1.2_{10}^{-8}	2.94	4.4_{10}^{-9}	3.69

It is striking that for the solution (5.4) the observed orders p_2 of DIRK2 and DIRK3 are in much better agreement with their orders p than for the solution (5.2). This indicates that for (5.4) the contamination of their local errors with large elementary differentials is much less than for (5.2) due to the zero boundary values. We again refer to Table 4.1 for comparison. Also note that for the larger τ_B-values DIRK2 and DIRK3 are hardly more efficient than MIDPOINT.

Problem III. The nonlinear problem

$$u_t = (u^5)_{xx},\ \ 0 < t \leqslant T = 1,\ \ 0 \leqslant x \leqslant 1, \tag{5.5}$$

discussed by Richtmyer & Morton [15], §8.6. They consider the running wave solution implicitly defined by $\frac{5}{4}(u-u_0)^4 + \frac{20}{3}u_0(u-u_0)^3 + 15u_0^2(u-u_0)^2 + 20u_0^3(u-u_0) + 5u_0^4\ln(u-u_0) = v(vt-x)$, where v, u_0 are constants. This is a wave running to the right if $v > 0$. Following [15], initial and (Dirichlet) boundary values were taken from (5.6) by Newton-Raphson solution. Results are given in Table 5.3 for $v = 10, u_0 = 1$.

Table 5.3. Results for the nonlinear problem (5.5).

τ_B	$EULER(\tau_B)$		$MIDPOINT(\tau_B)$		$DIRK2(2\tau_B)$		$DIRK3(3\tau_B)$	
	$\lVert\epsilon^N\rVert_2$	p_2	$\lVert\epsilon^N\rVert_2$	p_2	$\lVert\epsilon^N\rVert_2$	p_2	$\lVert\epsilon^N\rVert_2$	p_2
1/12	1.1_{10}^{-5}		3.9_{10}^{-3}		3.1_{10}^{-3}		4.4_{10}^{-3}	
1/24	5.4_{10}^{-6}	1.03	5.5_{10}^{-4}	2.82	3.6_{10}^{-4}	3.09	7.7_{10}^{-4}	2.50
1/48	2.7_{10}^{-6}	1.02	9.7_{10}^{-5}	2.50	7.8_{10}^{-5}	2.22	1.5_{10}^{-4}	2.40
1/96	1.3_{10}^{-6}	1.01	1.8_{10}^{-5}	2.44	1.9_{10}^{-5}	2.01	3.6_{10}^{-5}	2.02
1/192	6.7_{10}^{-7}	1.00	3.4_{10}^{-6}	2.39	4.8_{10}^{-6}	2.02	8.9_{10}^{-6}	2.03

Also for this problem DIRK2 and DIRK3 both suffer from a distinct order reduction. However, EULER and MIDPOINT behave uncommon, too. The observed order of MIDPOINT is clearly higher than two, while, not withstanding its order one, EULER yields remarkably accurate results. The explanation lies in the fact that u is non-smooth in the sense that higher derivatives of u are much larger than the lower ones (differentiate, e.g., the solution for $u_0 = 0$). In such situations EULER may operate more accurately than higher order schemes because for EULER the error depends essentially on the size of u_{tt}. The peculiar behaviour of MIDPOINT must be due to some lucky cancellation. Finally, the appearance of $p_2 = 2.0$ for DIRK2 and DIRK3 indicates that the reduction is dominated by a phenomenon as discussed in Example 4.2.

Our numerical experiments lead us to the following conclusions: (i) The experiments support our conjecture of §4 which states that the order q of DIRK2 and DIRK3 in the error bound (3.8) is at least 2. We proved this for semi-linear problems of the type $\dot{U} = AU + G(t, U)$ [3]. (ii) For many problems order reduction will decrease seriously the performance of DIRK2 and DIRK3. In case of time dependent boundary conditions the quantity p_2 given in (4.28) will be nearly equal for these two methods and close to the conjectured lower bound 2. (iii) DIRK2 and DIRK3 shall in general not perform better than MIDPOINT, neither in the high accuracy region due to order reduction. Our experiments strongly indicate that mostly the three schemes will be competitive to each other.

Acknowledgements This paper is a sequel to [18] which was written jointly with Prof. Chus Sanz-Serna from the University of Valladolid. With great pleasure I acknowledge many stimulating discussions with him. I also wish to thank Dr. Kevin Burrage from the University of Auckland and Dr. Willem Hundsdorfer for many helpful discussions on the order reduction phenomenon. Margreet Louter-Nool has taken care of the numerical experiments. She is to be acknowledged for her patience in doing a lot of trial and error testing.

References

1 Axelsson, O., Error estimates over infinite intervals of some discretizations of evolution equations, BIT 24, (1984) 413-424.

2 Brenner, P., M. Crouzeix & V. Thomée, Single step methods for inhomogeneous linear differential

equations in Banach space, R.A.I.R.O. Analyse numérique 16, (1982) 5-26.

3 Burrage, K., W.H. Hundsdorfer & J.G. Verwer, A study of B-convergence of Runge-Kutta methods, in press.

4 Cole, J.D., On a quasilinear parabolic equation occurring in aerodynamics, Quart. Appl. Math. 3, (1951) 225-236.

5 Crouzeix, M., Sur l'approximation des équations différentielle opérationelles linéaires parr des méthodes de Runge-Kutta, These, Université Paris VI, 1975.

6 Dahlquist, G., Stability and error bounds in the numerical integration of ordinary differential equations, Trans. Royal Inst. of Technology, No 130, Stockholm, 1959.

7 Dekker, K. & J.G. Verwer, Stability of Runge-Kutta methods for stiff nonlinear differential equations, North-Holland, Amsterdam-New York-Oxford, 1984.

8 Desoer, C. & H. Haneda, The measure of a matrix as a tool to analyze computer algorithms for circuit analyses, IEEE Trans. Circuit Theory 19, (1972) 480-486.

9 Frank, R., J. Schneid & C.W. Ueberhuber, The concept of B-convergence, SIAM J. Numer. Anal. 18, (1981) 753-780.

10 Frank, R., J. Schneid & C.W. Ueberhuber, Order results for implicit Runge-Kutta methods applied to stiff systems, Bericht Nr.53/82, Institut für Numerische Mathematik, TU Wien, 1982 (to appear in SIAM J. Numer. Anal.).

11 Kraaijevanger, H.F.B.M., B-convergence of the implicit midpoint rule and the trapezoidal rule, Report 01-1985, Inst. of Appl. Math. and Comp. Sc., University of Leiden, 1985.

12 Kreiss, H.O., Ueber die Stabilitätsdefinition für Differenzengleichungen die partielle Differentialgleichungen approximieren, BIT 2, (1962) 153-181.

13 Montijano, J.I., Estudio de los metodos SIRK para la resolucion numérica de ecuaciones differenciales de tipo stiff, Thesis, University of Zaragoza, 1983.

14 Norsett, S.P., Semi-explicit Runge-Kutta methods, Rep. Math. and Comp. No. 6/74, Dept. of Math., Univ. of Trondheim, 1974.

15 Richtmyer, R.D. & K.W. Morton, Difference methods for initial value problems, Interscience Publishers, New York-London-Sydney, 1967.

16 Stetter, H.J., Zur B-Konvergenz der impliziten Trapez-und Mittelpunktregel, Unpublished Note.

17 Varah, J.M., Stability restrictions on second order, three level finite difference schemes for parabolic equations, Technical Report 78-9, The University of British Columbia, Vancouver, 1978.

18 Verwer, J.G. & J.M. Sanz-Serna, Convergence of method of lines approximations to partial differential equations, Computing 33, (1984) 297-313.

19 Whitham, G.B., Linear and nonlinear waves, Wiley-Interscience, New York, 1974.

J.G. Verwer
Centre for Mathematics and Computer Science
Kruislaan 413, 1098 SJ Amsterdam
The Netherlands

W L WENDLAND

Splines versus trigonometric polynomials — the *h*- versus the *p*-version in two-dimensional boundary integral methods

1. INTRODUCTION

In [2], different finite element methods for solving the same problem have been compared systematically in order to make a somehow optimal choice. Here we try to pick this idea up for boundary element methods where even for one of the many possible governing boundary integral equations one has many different possibilities of their numerical treatment. However, for a systematical comparison one needs a reliable error analysis for all these methods which is currently available only for boundary integral methods applied to two-dimensional elliptic boundary value problems. We therefore restrict ourselves to two dimensions although in practical computations the three-dimensional problems are of much more interest. However, there the error analysis for the most common boundary element methods with collocation is not yet available. On the other hand, global approximation for three-dimensional problems is also extremely rare.

Here we first review asymptotic error estimates for splines with point collocation, with the Galerkin method and with the least squares method; global i.e. trigonometric polynomials with collocation, with Galerkin's method (which is a spectral method), with the least squares method using trigonometric polynomials and for Arnold's Galerkin-Petrov method with splines as trial and trigonometric polynomials as test functions. There asymptotic error estimates for the usual boundary element methods, i.e. the approximations with the "local" splines are formulated for decreasing meshwidth but fixed degree of the splines, i.e. for h-versions. The other extreme can be seen in the exponential convergence of global methods with trigonometric polynomials, which have the characteristics of spectral methods. From the comparison of both concepts we conclude that the most efficient method would be a combination of both worlds in a strategy similar to finite element p-versions. However, for such a strategy combining an adaptive mesh refinement controlled by a suitable error indicator with a systematical growth of the spline degrees, the mathematical foundation is almost completely missing - here I merely propose a concept. The corresponding mathematical justification is yet to be done.

2. THE BOUNDARY INTEGRAL EQUATIONS

As a typical situation for boundary integral methods let us consider second order elliptic boundary value problems in the plane. For simplicity let Ω_1 be a simply connected bounded domain having a smooth Jordan curve Γ as boundary and let Ω_2 denote the exterior complementary domain. We consider the regular elliptic boundary value problem in Ω_1 or Ω_2

$$\begin{aligned} P^{(2)}U &= 0 \quad \text{in } \Omega_j , \\ R\gamma_j U &= g \quad \text{on } \Gamma \end{aligned} \tag{2.1}$$

where U denotes the p-vector-valued unknown,

$$\gamma_j U = \binom{U}{\partial_n U} \quad \text{on } \Gamma \quad \text{for } U \text{ in } \Omega_j$$

and $R\gamma$ is a boundary operator of order $\mu = 0$ or 1. For $\mu = 0$, $R = (R_o,0)$ where R_o is a p×p coefficient matrix and for $\mu = 1$, $R = (R_o,R_1)$ with R_1 a coefficient matrix and R_o a tangential differential operator of order 1. ∂_n denotes the normal derivative with respect to the exterior normal direction to Ω_1 on Γ. For exterior problems in Ω_2 we also need appropriate radiation conditions at ∞. In the derivation of boundary integral equations to (2.1) we now follow the presentation in [11]. (See also [9, Chap. 5].) Let us assume that G is a fundamental solution to $P^{(2)}$, i.e. G represents the two-sided inverse to $P^{(2)}$ on the distributions with compact support in $\mathbb{R}^2$. For the boundary conditions $R\gamma_j$ let us assume that we can find complementing boundary conditions $S\gamma_j$ such that the matrix $M = \binom{R}{S}$ is invertible along Γ. Further let us define along Γ the tangential differential operators P_o,P_1,P_2 by the decomposition

$$P^{(2)}|_\Gamma = P_o + P_1\partial_n + P_2\partial_n^2 .$$

Then any solution U of the boundary value problem (2.1) can be expressed by Green's representation formula

$$\begin{aligned} U(x) = (-1)^j \int_\Gamma \{ &(-P'_{1(y)}G(x,y) + P_2\partial'_{n(y)}G(x,y))U(y) \\ &+ P_2G(x,y)\partial_n U(y) \} ds_y + W(x)\omega \quad \text{for } x \in \Omega_j . \end{aligned} \tag{2.2}$$

Here $W(x)\omega$ denotes a suitable linear combination of solutions to $P^{(2)}W\omega = 0$ with linear factors $\omega \in \mathbb{R}^q$ which take care of the radiation at ∞ for exterior problems [20], but which might be neglected for interior problems. P_1' and ∂_n' denote the adjoint differential operators. According to the boundary condition let us introduce yet unknown boundary functions v via

$$\binom{U}{\partial_n U}\Big|_\Gamma = \binom{R}{S}^{-1}\binom{g}{v} = M^{-1}\binom{g}{v} ,$$

insert $\gamma_j U$ into (2.2) and obtain on the boundary

$$\binom{g(x)}{v(x)} = \binom{R}{S}\gamma_j U = (-1)^j \binom{R}{S}\gamma_j \int_\Gamma \{(-P_1'{}_{(y)} + P_2\partial_n'{}_{(y)})G(x,y),\ P_2 G(x,y)\} \cdot$$

$$\cdot \binom{R}{S}^{-1}\binom{g(y)}{v(y)}\, ds_y + \binom{R}{S}\gamma_j W\omega \quad \text{for } x \in \Gamma .$$

This is an overdetermined system of boundary integral equations for v and one still has different possibilities to choose a quadratic subsystem which is equivalent to the original one. When this choice has been taken we find a boundary integral equation of the form

$$\begin{aligned} Av(x) + B(x)\omega &= f(x) \qquad \text{for } x \in \Gamma , \\ \Lambda v &= \beta \end{aligned} \tag{2.3}$$

with appended equilibrium conditions, where v denotes the desired boundary charge, a p-vector valued function. $\omega \in \mathbb{R}^q$ are unknown constants. $f(x)$ and $\beta \in \mathbb{R}^q$ are given, $B(x)$ is a given $p\times q$ matrix of functions. Λ is a given $q\times p$ matrix of linear functionals and A is an elliptic (matrix) pseudo-differential operator on Γ of integer order $2\alpha \in \mathbb{Z}$ [11]. Let us consider more explicitly the following two examples for (2.3) .

2.1 Logarithmic Fredholm integral equations of the first kind

Let $y = z(\sigma)$, $x = z(s)$ denote a 1-periodic regular representation of the curve Γ . In many applications, equations (2.3) take the form of Fredholm integral equations of the first kind as

$$Av(s) = -\frac{1}{\pi}\int_{\sigma=0}^{1} (\log|z(s)-z(\sigma)| - L(s,\sigma))v(\sigma)d\sigma \tag{2.4}$$

with a smooth kernel L , i.e. A is a pseudo-differential operator of order $2\alpha = -1$ having the principal symbol $a_0(s,\xi) = |\xi|^{-1}$. Hence A in (2.4) is even strongly elliptic (see Chap. 4). Equations (2.3) with operators of the form (2.4) can be met in many applications as electrostatics, acoustics, elasticity, viscous flow problems and conformal mappings. For many corresponding references see [17, 18, 36].

2.2 Integro-differential equations

In the same applications as mentioned above we also find Cauchy singular boundary integral equations, Fredholm integral equations of the second kind, ordinary and integro-differential equations including some with hypersingular non integrable kernels which can all be written in the form [4]

$$\begin{aligned} Av(s) &= A_o(s)\frac{d^\ell v}{ds^\ell} + \sum_{j=0}^{\ell-1} A_{\ell-j}\frac{d^j v}{ds^j} \\ &\quad + \frac{1}{\pi i}\int_{\sigma=0}^{1}\{\sum_{j=0}^{\ell} C_{\ell-j}(s,\sigma)\frac{d^j v}{d\sigma^j}(\sigma)\}\frac{d\zeta}{\zeta-z(s)} + \int_0^1 L(s,\sigma)v d\sigma \end{aligned} \tag{2.5}$$

where $\zeta = y_1 + iy_2$, $z = x_1 + ix_2$. Here the order of A is $2\alpha = \ell \in \mathbb{N}_o$ and the principal symbol is given by

$$a_o(s,\xi) = i^\ell \xi^\ell (A_o(s) + C_o(s,s)\frac{\xi}{|\xi|}) .$$

3. TRIAL AND TEST FUNCTIONS

As approximating function spaces for our 1-periodic functions we consider either global trigonometric polynomials or "smoothest" splines. These spaces will serve as trial functions and as test functions. Let N and d denote integers. For given N let us denote by T_N the span of trigonometric polynomials

$$\begin{aligned} \tau_{j+1} &= \text{Re}\exp(2\pi ijs) , \quad j = 0,\dots,[\tfrac{N}{2}] \text{ and} \\ \tau_{j+[\frac{N}{2}+1]} &= \text{Im}\exp(2\pi ijs) , \; j = 1,\dots,[\tfrac{N}{2}] , \end{aligned} \tag{3.1}$$

$[\cdot]$ denotes the Gaussian bracket. For the "smoothest" splines we follow [6] defining first the reference functions

$$\tilde{\mu}^o(\xi) = \begin{cases} 1 & \text{for } 0 \le \xi < 1 \\ 0 & \text{otherwise} \end{cases}$$

and then $\tilde{\mu}^d(\xi)$ recursively by the convolutions

$$\tilde{\mu}^{d+1}(\xi) = \int_{\mathbb{R}} \tilde{\mu}^d(\xi-\eta)\tilde{\mu}^o(y)d\eta , \quad d = 0,1,\dots .$$

For the bases spanning the spline spaces S_N^d we choose for fixed d and N ,

$$\mu_j(s) := \tilde{\mu}^d\left(\frac{s}{h} - j+1\right) \quad \text{for} \quad h\cdot j \le s+h \le 1 + h\cdot j \ , \tag{3.2}$$

where $j = 1,2,\ldots,N$ and $h = N^{-1}$ and for arbitrary s their 1-periodic extensions. Then S_N^d consists of all (d-1)-times continuously differentiable 1-periodic functions defined by piecewise polynomials of degree d subordinate to the partition $\{j\cdot h|\ j\in \mathbb{Z}\}$.

For the formulation of approximation and convergence properties we shall use the Sobolev spaces H^r of 1-periodic functions of arbitrary real order r , i.e. the closure of all smooth 1-periodic functions with respect to the norm

$$\| f \|_r = \langle f,f\rangle_r^{\frac{1}{2}}$$

induced by the scalar product

$$\langle f,g\rangle_r := \sum_{k\in\mathbb{Z}} \hat{f}_k \overline{\hat{g}_k} |2\pi k + \delta_{ok}|^{2r}$$

where

$$\hat{f}_k = \int_0^1 f(s)\exp(-2\pi kis)ds \ , \ k \in \mathbb{Z}$$

are the complex Fourier coefficients.

If P_N denotes the L^2-orthogonal projection onto T_N then the trigonometric polynomials provide the

Global approximation property (Theorem by Jackson) [15, 22].

Let $-\infty < r \le \rho$. *Then for* $f \in H^\rho$ *we have*

$$\| P_N f-f \|_r \le c_1 N^{r-\rho} \| f \|_\rho \quad .$$

If f *is real analytic having a complex analytic extension to* $S_a = \{z \in C|\ |\mathrm{Im}\, z| \le a\}$ *with* $a > 0$ *then*

$$\| P_N f-f \|_r \le c_2 N^r e^{-2\pi aN} \sup_{z\in S_a} |f(z)| .$$

The constants c_1, c_2 *are independent of* N *and* f *and* c_2 *also of* r *and* a .

T_N also provides the

Inverse assumption [22].

For every $r \le \rho$ *there holds*

$$\| \tau \|_{\rho} \le cN^{\rho - r} \| \tau \|_{r} \qquad \text{for all} \quad \tau \in T_{N}$$

where c *is independent of* τ *and* N .

As is well known, the spline spaces S_N^d provide similar properties.

Global approximation property [1, 4, 12, 14].

Let $\rho' < d + \frac{1}{2}$. *Then there exists a family of approximation operators* $P_h^d : H^{\rho'} \to S_N^d$ *such that if* $r \le \rho' \le \rho \le d+1$ *then*

$$\| P_h^d f - f \|_{r} \le ch^{\rho - r} \| f \|_{\rho} , \qquad f \in H^{\rho}$$

where c *does not depend on* h *or* f .

Inverse property [12].

Let $r \le \rho < d + \frac{1}{2}$. *Then there exists a constant* c *such that*

$$\| \chi \|_{\rho} \le ch^{r - \rho} \| \chi \|_{r} \qquad \text{for all} \quad \chi \in S_N^d .$$

However, in addition, splines S_N^d also provide a

Local approximation property [24, 25].

Let $I_o = [\ell_o, r_o]$, $I_1 = [\ell_1, r_1]$ *be two intervals with* $\ell_1 < \ell_o < r_o < r_1$ *and let* $r \le \rho' \le \rho$ *and* P_h^d *be as in the global approximation property. Then*

$$\| P_h^d f - f \|_{H^r(I_o)} \le ch^{\rho - r} \{ \| f \|_{H^{\rho}(I_1)} + \| f \|_{r} \} \tag{3.3}$$

where c *does not depend on* h *or* f .

The local norms are defined by

$$\| f \|_{H^r(I)} = \inf_{f = \phi|_I} \| \phi \|_{r} .$$

Note that the local approximation property holds for splines but *not for global polynomials* [24] .

4. SOME BOUNDARY ELEMENT AND BOUNDARY SPECTRAL METHODS AND THEIR CONVERGENCE

Here we shall review some of the known convergence properties for a whole variety of different methods for solving boundary integral equations (2.3) numerically. They all can be formulated in an unifying manner in terms of Galerkin-Petrov methods, where trial and test functions can be different.

Let $\{u_j\}_{j=1}^N$ be a basis of the trial space, then

$$v_h = \sum_{j=1}^N \underset{\sim}{\gamma}_j u_j(s) \quad \text{and} \quad \omega_h$$

denote the desired approximate solutions of (2.3). Let $\{T_k\}_{k=1}^N$ denote a basis of some test space. Then the Galerkin-Petrov method reads as:

Find $\underset{\sim}{\gamma}_j \in \mathbb{R}^p$ *and* $\omega_h \in \mathbb{R}^q$, $j = 1,\ldots,N$ *by solving the quadratic algebraic linear equations*

$$\sum_{j=1}^N \underset{\sim}{\gamma}_j (Au_j, T_k)_{L^2(\Gamma)} + (B\omega_h, T_k)_{L^2(\Gamma)} = (f, T_k)_{L^2(\Gamma)} ,$$

$$\sum_{j=1}^N \underset{\sim}{\gamma}_j \Lambda u_j = \beta , \quad k = 1,\ldots,N .$$

The best known and most common choices are listed in the following Table 1.

Method	Trial	Test
1.) Boundary element methods as naive spline collocation [8,38]	μ_j	$\delta(s-h(k+\frac{d+1}{2}))$
2.) Boundary element Galerkin method with splines [19]	μ_k	μ_k
3.) Spline least squares [21,37]	μ_j	$A\mu_k$
4.) Spline-trig method [1], N odd	μ_j	τ_k
5.) Trigonometric collocation [28]	τ_j	$\delta(s-hk)$
6.) Spectral method [13, 15]	τ_j	τ_k
7.) Least squares spectral method [21]	τ_j	$A\tau_k$

Table 1. Trial and test apces for different methods.

For the above methods we are able to compare the convergence properties.

However, it should be noted that the integrations in (4.1) must be performed numerically and that the corresponding errors will produce additional errors in the computational results. The latter will be analyzed in the next section.

The global asymptotic error estimates for the above methods are of the form

$$\| v_h - v \|_r + |\omega_h - \omega| \leq ch^{\rho - r} \| v \|_\rho \tag{4.2}$$

or

$$\| v_h - v \|_r + |\omega_h - \omega| \leq c_1 e^{-c_2 N} \sup_{z \in S_a} |v(z)| \tag{4.3}$$

where the positive constants c, c_1, c_2 are independent of h, v, v_h. The methods 3.), 5.), 6.), 7.) converge already for elliptic problems whereas for the convergence of the methods 1.), 2.) and 4.) we need for the operator A the more restrictive assumption of <u>strong ellipticity</u>: *To the principal symbol* a_o *of* A *there exist a positive constant* γ *and a real* $p \times p$ *1-periodic* C^∞ *matrix-valued function* $\Theta(s)$ *such that*

$$\mathrm{Re}\ \zeta^T \Theta(s) a_o(s,\xi) \bar{\zeta} \geq \gamma |\zeta|^2$$

for all s , *all* $\zeta \in \mathbb{C}^p$ *and* $\xi = \pm 1$. (For another characterization see [26].)

The corresponding results and references are listed in Table 2. In the last columns we also list the ℓ_2-ℓ_2 inverse stability with $\alpha' := \min\{0,\alpha\}$ and conditioning of the discrete equations (4.1), provided ω_h and β are optimally scaled (see [10]). For 1.) we only consider naive collocation and the results given in [4, 5, 30], but is hould be mentioned that in [27] and [32] one finds further results and references as well as modifications of spline collocation methods which work for a larger class than the strongly elliptic operators. The results for 5.) and 7.) can be found or be easily obtained from [28] in the case of Cauchy singular integral equations, $2\alpha = 0$. For $2\alpha \neq 0$ use the modified differential operator D [5, (5.4)] and take $D^{2\alpha} T_N$ instead of T_N to reduce this case to the case $2\alpha = 0$ as in [22]. Inverse stability and conditioning are also obtained as in [22].

For the spline-trig method 4.), D.N. Arnold introduces in [1] periodic Hilbert spaces $H_{r,\varepsilon}$ for $\varepsilon > 0$ subordinate to

Method	Restrictions for Sobolev indices and references for proofs	inverse stability	conditioning
1.)	(4.2) with $2\alpha \le r \le \rho \le d+1$; $r, 2\alpha < [\frac{d}{2}] + \frac{d+1}{2}$, [4, 5, 30, 31]	$N^{-2\alpha'}$	$N^{\|2\alpha\|}$
2.)	(4.2) with $2\alpha-d-1 \le r \le \rho \le d+1$; $r, \alpha < d + \frac{1}{2}$, [4, 19, 31]	$N^{-2\alpha'}$	$N^{\|2\alpha\|}$
3.)	(4.2) with $4\alpha-d-1 \le r \le \rho \le d+1$; $r, 2\alpha < d + \frac{1}{2}$, [31, 33]	$N^{-4\alpha'}$	$N^{\|4\alpha\|}$
4.)	(4.4) with $-\infty < r \le \rho \le d+1$; $\alpha, r < d + \frac{1}{2}$, ex. $\varepsilon<1$, [1]	$N^{-2\alpha'}$	$N^{\|2\alpha\|}$
5.)	(4.2) with $-\infty < r \le \rho < \infty$ and (4,3) , [28]	$N^{-2\alpha'}$	$N^{\|2\alpha\|}$
6.)	(4.2) with $-\infty < r \le \rho < \infty$ and (4.3) , [13, 22, 23, 28]	$N^{-2\alpha'}$	$N^{\|2\alpha\|}$
7.)	(4.2) with $-\infty < r \le \rho < \infty$ and (4.3) , [13]	$N^{-4\alpha'}$	$N^{\|4\alpha\|}$

Table 2. Asymptotic orders of convergence, stability and conditioning.

$$\langle f,g\rangle_{r,\varepsilon} = \sum_{k\in\mathbb{Z}} \hat{f}_k \overline{\hat{g}_k} \, |2k\pi + \delta_{ok}|^{2r} \varepsilon^{2|k|} .$$

Here the error estimate reads as

$$\| v_h - v \|_{r,\varepsilon} + |\omega_h - \omega| \le c \, \varepsilon^{\frac{N}{2}} N^{r-\rho} \| v \|_\rho \qquad (4.4)$$

In this case as well as for (4.3) one needs real analytic data and Γ .

5. FULL DISCRETIZATION WITH NUMERICAL INTEGRATION

For the numerical execution of any of the above methods we need to compute the weights in (4.1), i.e.

$$a_{jk} = (Au_j, T_k) , \qquad f_k = (f, T_k) ,$$

by numerical integrations. If we require that their computed values $\tilde{a}_{jk}$ and $\tilde{f}_k$ are precise in the sense of satisfying

$$|\tilde{a}_{jk} - a_{jk}| + |\tilde{f}_k - f_k|h \le ch^{\sigma} \tag{5.1}$$

where c is independent of j,k and h then with the inverse stability as given in the preceding section together with the inverse assumption and approximation properties, the second Strang lemma implies that the additional errors can be estimated as

$$\| \tilde{v}_h - v_h \|_r + |\tilde{\omega}_h - \omega_h| \le c(v,f)h^{\tau} \tag{5.2}$$

where $\tau > 0$ is related to σ, r and α due to the following Table 3. (See [18, 35])

Method :	1.)	2.)	3.)	4.)	5.)	6.)	7.)
$\tau = \sigma - r' + 2\alpha' +$	-1	-2	$-2 + 2\alpha'$	-1	-1	-1	$-1 + 2\alpha'$

Table 3. Precision and order of convergence.

Hence, the highest optimal orders of convergence in Table 2 for the different methods is recovered by the fully discretized scheme if σ in (5.1) is chosen at least as in Table 4.

Method :	1.)	2.)	3.)	4.)	5.) 6.)	7.)
$d+2-2\alpha' +$	-2α	$d+2-2\alpha$	$d+2-4\alpha-2\alpha'$	$-\frac{N}{2}\frac{\log \varepsilon}{\log N}$	$\frac{N}{\log N} - d-1$	$\frac{N}{\log N} - 2\alpha' - d-1$

Table 4. Least orders of σ for τ corresponding to the highest orders in Table 2.

Table 4 shows that for the methods 1.) - 3.) for fixed d the precision of integration σ in (5.1) is indpendent of the number of grid points N, i.e. on every <u>patch</u> one may use the <u>same</u> numerical quadrature rule. In order to pick up the exponential convergence of the methods 4.) - 7.), however, one now needs a recursive adaption of numerical cubature due to the dependence of σ on N.

In order to perform a p-method, i.e. increase d for fixed N or for slowly increasing N it is advantageous to use numerical quadratures with the regular grid points $\{h(k+\tilde{\gamma})\}$ with $\tilde{\gamma} = 0$ and, perhaps $\tilde{\gamma} = \frac{1}{2}$ and $\frac{1}{4}$ etc. Grid point integrations of this type have been developed in [3, 17, 18, 34, 37] for operators of the form (2.4) and (2.5). These take always the form ((3.5) after integration by parts) of sums of operators

$$A_\lambda v(s) = a_\lambda(s) \frac{d^\lambda v(s)}{ds^\lambda} + c_\lambda(s) \int_{t=0}^{1} \log|s-t| \frac{d^{\lambda+1} v}{dt^{\lambda+1}}(t)\, dt + \int_{t=0}^{1} L_\lambda(s,t)v(t)\, dt \,. \tag{5.3}$$

Hence, for $v = \mu_j$ given by (3.2), the second term in (5.3) becomes for spline collocation 1.)

$$h^{-\lambda} c_\lambda(h(k + \frac{d+1}{2})) \int_0^{d+1} [\log h + \log|t' + (j-k) - \frac{d+1}{2}|] \frac{d^{(\lambda+1)}\tilde{\mu}^d}{(dt')^{\lambda+1}}(t')dt' \,. \tag{5.4}$$

Clearly, the integrals in (5.4) can be computed explicitly by using either the analytic formula or a sufficiently high order weighted Gaussian quadrature. Note, that the second integral in (5.4) depends only on $(j-k)$, λ and $\tilde{\mu}^d$ and could be tabelized. The last remaining integral in (5.3) can be treated by the grid point rule,

$$\int_{t=0}^{1} L_\lambda(h(k+\frac{d+1}{2}),t)\mu_j(t)dt = \sum_{\ell=-M}^{M} b_\ell L_\lambda(h(k+\frac{d+1}{2}),\ h(j+\frac{d+1}{2}+\ell\tilde{\gamma})) + O(h^{2M+3}) \tag{5.5}$$

where the weights $b_\ell = b_{-\ell}$, $\ell = 0,\dots,M$ are defined by the moment fitting equations

$$\int_0^1 \mu_j(t)F(t)dt = h \sum_{\ell=-M}^{M} b_\ell F(h(j+\frac{d+1}{2} + \tilde{\gamma}\ell)) \tag{5.6}$$

with $\tilde{\gamma} = 1$ (or $\frac{1}{2}$ or $\frac{1}{4}$ if necessary) for all polynomials F of degree up to $2M+1$ (see [3, 17, 18, 34, 37]). Note, here

$$\sigma = 2M + 3 \,.$$

Now the transition in the method 1.) from d to $d+1$ will require new weights b_ℓ (which in fact are very simple and known) whereas the function values of L_λ in (5.5) can be stored and be used again saving costly re-evaluations.

For Galerkin's method with splines we proceed correspondingly obtaining for the terms in a_{jk} coming from (5.3)

$$h^{1-\lambda}\int_0^{d+1}\tilde{\mu}^d(\tau')\,\{a_\lambda(h(\tau'+k+\frac{d+1}{2}))((\frac{d}{d\tau'})^\lambda\tilde{\mu}^d(\tau'))+c_\lambda(h(\tau'+k+\frac{d+1}{2}))$$

$$\cdot\int_0^{d+1}[\log h+\log|t'-\tau'+(j-k)|]((\frac{d}{dt'})^{(\lambda+1)}\tilde{\mu}^d(t'))dt'\}d\tau'$$

$$+\,h^2\sum_{m,\ell=-M}^{M} b_\ell b_m L_\lambda(h(k+\frac{d+1}{2}+m\tilde{\gamma}),\,h(j+\frac{d+1}{2}+\ell\tilde{\gamma}))\,. \qquad (5.7)$$

For the remaining integral with weight $\tilde{\mu}^d(t')$ we may use again the grid point rule (5.6) in combination with exact integration of the inner logarithmic integral as in (5.4). For constant c_λ and for $\lambda = -1$ the double integrals in (5.7) have been evaluated exactly and tabelized in [18, Table 1.3.1]. In both cases, here we have the precision

$$\sigma = 2M + 4$$

in (5.1). Note that these formulas lead to a modified collocation, the so called *Galerkin collocation* which also allows the transition from d to $d+1$ without changing the grid point values of the kernel L_λ.

For the methods involving the global trigonometric polynomials note that inserting these into the pseudo-differential operator A having symbol $a(s,\xi)$ takes the explicit form

$$Ae^{2\pi ik\cdot}(s) = a(s,k)e^{2\pi iks}\,.$$

In particular for the principal part with principal symbol a_o we have for $k \neq 0$

$$A_o e^{2\pi ik\cdot}(s) = a_o(s,\frac{k}{|k|})\,|k|^{2\alpha}\,e^{2\pi iks}$$

and numerical integration is needed only for the Fourier coefficients of $a_o(s,\frac{k}{|k|})$ for fixed k and of the remaining kernels. Here use a family of uniform subdivisions of the unit interval into

$$M = 2^{\tilde{q}} \geq N$$

subintervals of length M^{-1} in combination with fast Fourier transform for the integrals. This yields a very efficient spectral method proposed and analyzed in [15]. In [22] the repeated use of a fixed Gaussian quadrature formula on each subinterval in combination with the FFT proved to be an extremely fast and accurate method. If the functions involved are real analytic, then (5.1) holds with any σ and one even has

$$|\tilde{a}_{jk} - a_{jk}| + |\tilde{f}_k - f_k| h \leq c_1 e^{-c_2 N}$$

which implies the requirements in Table 4 and exponential convergence as in Table 2 also for the fully discretized versions of methods 5.) - 7.).

6. , OBESERVATIONS AND CONCLUSIONS

The asymptotic convergence rates of the previous chapters indeed are revealed in numerical experiments. In Figure 1 we present experiments for solving Symm's integral equations of the type (2.4) for conformal mapping of reflected ellipses from [16] with spline collocation, [18] with Galerkin collocation and [22] with the spectral method combined with FFT . The spectral method not only converges fastest but also gives the most accurate results with least degrees of freedom. However, it requires extremely smooth data *everywhere* on Γ and also for Γ . This is in practical applications often violated and local nonsmoothness pollutes the results everywhere. This is in contrary to the splines which not only provide the local approximation property (3.3), but at least the spline Galerkin method also converges locally with highest possible orders, i.e., local super approximation which has been proved in [7] :

For the spline Galerkin method 2.) *one has local super approximation*

$$\| u_h - u \|_{H^r(I_0)} \leq ch^{\rho - r}\{ \| v \|_{H^\rho(I_1)} + \| v \|_r \}$$

with $r \leq \rho$ *satisfying the restrictions in* Table 2, 2.). Moreover, *increasing* d *and doubling* h^{-1} *at the same time* yields a behaviour of spline methods as of the spectral method as can be seen from the spline collocations in Figure 1 , curve —·— . This indicates to give up globally uniform grids and to use uniform grids only on subintervals in combination

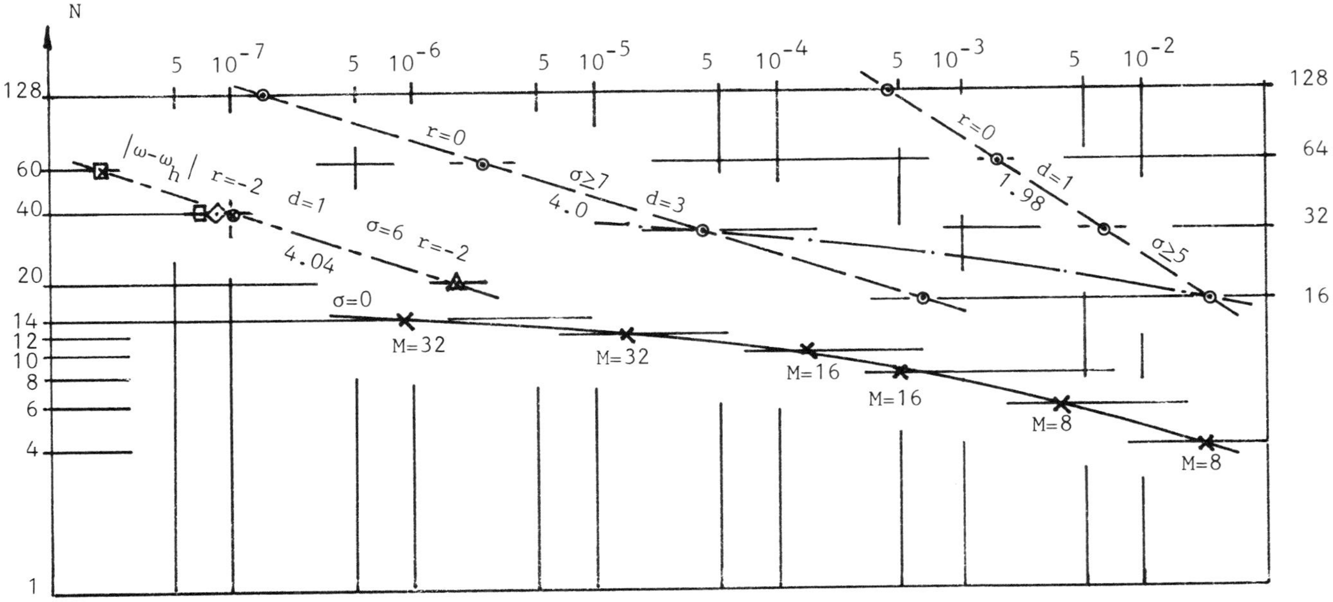

— — Spline collocation, smoothest splines [16]

—·— Spline Galerkin-collocation, smoothest splines [18]

—— Spectral method with FFT [22]

Figure 1. Errors for Symm's integral equation of the first kind with logarithmic kernel, $2\alpha = 1$.

with an adaptive meshrefinement. The following strategy seems to be rather promising (see Figure 2).

1. *Extend the error indicator developed in* [29] *to general strongly elliptic boundary integral equations.*
2. *Use an adaptive meshrefinement which is uniform on subregions.*
3. *Increase* d *to* d+1 .

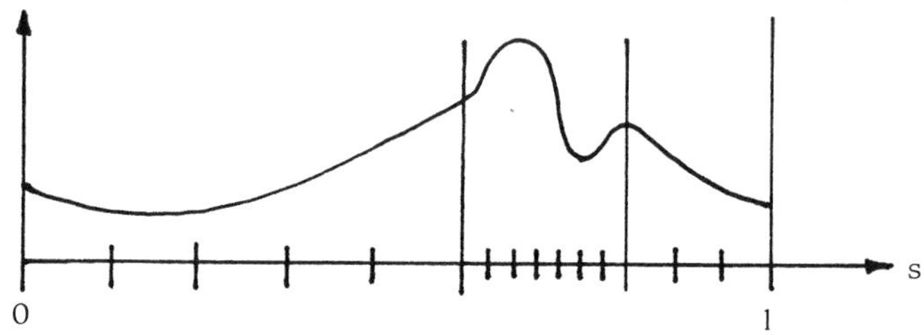

Figure 2.

As we have indicated in the previous chapter, with appropriate numerical quadrature rules the transition from d to d+1 is possible by computing and storing appropriate function values of kernels and right hand sides only at additional new grid points and reusing the saved values from the previous steps.

In view of our error estimates, such a strategy should produce extremely accurate BEM solutions within few steps.

On the other hand we are facing several open problems as to incorporate the meshrefinement and error indicator as well as the transition to higher degree splines in a simple manner into the structure of BEM codes; to find simple indicators which work; to analyze asymptotic error bounds exhibiting the dependence on h and d also for nonuniform grids explicitly (- the constants in the certified estimates (4.2) do depend on d -) and to perform experiments for finding out whether this strategy makes sense at all.

REFERENCES

1. ARNOLD, D.N. A spline-trigonometric Galerkin method and an exponentially convergent boundary integral method, Math. Comput. 41 (1983) 383-397.

2. ARNOLD, D.N., BABUŠKA, I. and OSBORN, J. Finite element methods: Principles for their selection, Comp. Meth. Appl. Mech. Eng. 45 (1984) 57-96.

3. ARNOLD, D.N. and WENDLAND, W.L. Collocation versus Galerkin procedures for boundary integral methods, in Boundary Element Methods in Engineering (C.A. Brebbia ed.), Springer-Verlag, Berlin 1982, 18-33.

4. ARNOLD, D.N. and WENDLAND, W.L. On the asymptotic convergence of collocation methods, Math. Comput. 41 (1983) 349-381.

5. ARNOLD, D.N. and WENDLAND, W.L. The convergence of spline collocation for strongly elliptic equations on curves, Numer. Math. (1985), in press.

6. AUBIN, J.P. Approximation of Elliptic Boundary-Value Problems. Wiley-Intersc., New York, 1972.

7. BRAUN, K. Lokale Konvergenz der Ritz-Projektion auf finite Elemente für Pseudo-Differentialgleichungen negativer Ordnung, Doctoral Thesis, Univ. Freiburg, Germany, 1985.

8. BREBBIA, C.A., TELLES, J.C.F. and WROBEL, L.C. Boundary Element Techniques, Springer-Verlag, Berlin, Heidelberg, New York, 1984.

9. CHAZARAIN, J. and PIRIOU, A. Introduction to the Theory of Linear Partial Differential Equations, North-Holland Publ., Amsterdam, New York, Oxford, 1982.

10. CHRISTIANSEN, S. and WENDLAND, W.L. On the condition number of the influence matrix belonging to some first kind integral equations with logarithmic kernel, Appl. Anal., in press.

11. COSTABEL, M. and WENDLAND, W.L. Strong ellipticity of boundary integral operators, to appear.

12. ELSCHNER, J. and SCHMIDT, G. On Spline Interpolation in Periodic Sobolev Spaces, P-Math.-01/83, Inst. f. Math., Akad. d. Wiss. DDR, Berlin 1983.

13. GOHBERG, I.C. and FEL'DMAN, I.A. Convolution Equations and Projection Methods for their Solution, Amer. Math. Soc., Providence, Rhode Island, 1974.

14. HELFRICH, H.-P. Simultaneous approximation in negative norms of arbitrary order, R.A.I.R.O. Numer. Anal. 15 (1981) 231-235.

15. HENRICI, P. Fast Fourier methods in computational complex analysis, SIAM Rev. 21 (1979) 481-527.

16. HOIDN, H.P. Die Kollokationsmethode angewandt auf die Symmsche Integralgleichung, Doctoral Thesis, ETH Zürich, Switzerland, 1983.

17. HSIAO, G.C., KOPP, P. and WENDLAND, W.L. A Galerkin collocation method for some integral equations of the first kind, Computing 25 (1980) 89-130.

18. HSIAO, G.C., KOPP, P. and WENDLAND, W.L. Some applications of a Galerkin collocation method for integral equations of the first kind, Math. Meth. Appl. Sci. 6 (1984) 280-325.

19. HSIAO, G.C. and WENDLAND, W.L. The Aubin-Nitsche lemma for integral equations, J. Integral Equations 3 (1981) 299-315.

20. HSIAO, G.C. and WENDLAND, W.L. On a boundary integral method for some exterior problems in elasticity. Dokl. Akad. Nauk SSR, to appear (Preprint 769, Fachber. Math. Techn. Hochschule Darmstadt 1983).

21. IVANOV, V.V. The Theory of Approximate Methods and their Application to the Numerical Solution of Singular Integral Equations, Noordhoff Intern. Publ., Leyden, 1976.

22. LAMP, U., SCHLEICHER, K.-T. and WENDLAND, W.L. The fast Fourier transform and the numerical solution of one-dimensional boundary integral equations, Numer. Math. 47 (1985) 15-38.

23. MC LEAN, W. Boundary Integral Methods for the Laplace Equation, Doctoral Thesis, Australian Nat. Univ., Canberra, Australia, 1985.

24. NITSCHE, J. and SCHATZ, A. On local approximation properties of L_2-projections on spline subspaces, Appl. Anal. 2 (1972) 161-168.

25. NITSCHE, J. and SCHATZ, A. Interior estimates for Ritz-Galerkin methods, Math. Comput. 28 (1974) 937-958.

26. PRÖSSDORF, S., Starke Elliptizität singulärer Integraloperatoren und Spline-Approximation, in Linear and Complex Analysis Problem Book - 199 Research Problems (U.P. Havin, S.V. Hruscëv, N.K. Nikolśki eds.), Lecture Notes in Math. 1043, Springer-Verlag, Berlin, Heidelberg, 1984.

27. PRÖSSDORF, S. and RATHSFELD, A. On spline Galerkin methods for singular integral equations with piecewise continuous coefficients, Numer. Math., to appear.

28. PRÖSSDORF, S. and SILBERMANN, B. Projektionsverfahren und die näherungsweise Lösung singulärer Gleichungen, B.G. Teubner Verl., Leipzig, 1977.

29. RANK, E. Adaptivity and accuracy estimation for finite element and boundary integral element methods, in Accuracy Estimation and Adaptivity for Finite Elements (I. Babuska, O.S. Zienkiewicz, Arantes e Oliveira eds.), Wiley-Intersc., New York, 1984.

30. SARANEN, J. and WENDLAND, W.L. On the asymptotic convergence of collocation methods with spline functions of even degree, Math. Comput. 45 (1985), in press.

31. SCHMIDT, G. The convergence of Galerkin and collocation methods with splines for pseudo-differential equations on closed curves, Zeitschr. Analysis u. i. Anwendungen 3 (1984) 371-384.

32. SCHMIDT, G. On spline collocation methods for boundary integral equations in the plane, Math. Meth. Appl. Sci. 7 (1985), to appear.

33. STEPHAN, E. and WENDLAND, W.L. Remarks to Galerkin and least squares methods with finite elements for general elliptic problems, Manuscripta Geodetica 1 (1976) 93-123.

34. WENDLAND, W.L. On Galerkin collocation methods for integral equations of elliptic boundary value problems, in Numerical Treatment of Integral Equations (J. Albrecht, L. Collatz eds.) ISNM, Birkhäuser, Basel 53 (1980) 244-275.

35. WENDLAND, W.L. Asymptotic accuracy and convergence, in Progress in Boundary Elements.(C.A. Brebbia ed.) Pentech Press, London, 1981, 289-313.

36. WENDLAND, W.L. Boundary element methods and their asymptotic convergence, in Theoretical Acoustics and Numerical Techniques (P. Filippi ed.) CISM Courses 277, Springer-Verlag, Wien, New York 1983, 135-216.

37. WENDLAND, W.L. On some mathematical aspects of boundary element methods for elliptic problems, in The Mathematics of Finite Elements and Applications V (J.R. Whiteman ed.) Academic Press, London, 1985, 193-227.

38. WENDLAND, W.L. Asymptotic accuracy and convergence for point collocation methods, in Topics in Boundary Element Research 2 (C.A. Brebbia ed.) Springer-Verlag, Berlin, 1985, 230-257.

W.L. Wendland
Technische Hochschule Darmstadt
Fachbereich Mathematik
Schlossgartenstrasse 7
6100 Darmstadt
West Germany

SUBMITTED PAPERS

C A Addison, Department of Computing Science, University of Alberta, Edmonton, Alberta, Canada T6G 2H1.
The implementation of hybrid formulae for initial value differential equations.

J M Aitchison, Royal Military College of Science, School of Mathematics and Management, Shrivenham, Swindon, Wiltshire, SN6 8LA, UK.
A boundary integral technique for the calculation of immiscible flow in a porous medium.

G Alefeld, Institut fur Angewandte Mathematik, Universitat Karlsruhe, Postfach 6380, Kaiserstrasse 12, 7500 Karlsruhe 1, W Germany.
Rigorous error bounds for singular values of a matrix using the precise scalar product.

P Aston* and J R Whiteman, Institute for Numerical Mathematics, Brunel University, Kingston Lane, Uxbridge, Middlesex, UB8 3PN, UK.
Finite element analysis for bifurcation of thin, shallow, spherical shells.

N F Attia* and C G Broyden, Department of Computer Science, University of Essex, Colchester, Essex, UK.
The simple quadratic penalty function and the S.Q.P. Methods.

L Baart, NRIMS-CSIR, P.O. Box 395, Pretoria 0001, South Africa.
Quadratic parametric transformations.

M J Baines, Mathematics Department, University of Reading, Whiteknights, P.O. Box 220, Reading, RG6 2AX, UK.
Local moving finite elements.

J W Barrett* and C M Elliott, Mathematics Department, Imperial College, Queens Gate, London, SW7 2BX, UK.
Finite element approximation of the Dirichlet problem using the boundary penalty method.

R H Bartels and N Mahdavi-Amiri*, Department of Computer Science, York University, 4700 Keele Street, Downsville, Ontario, Canada M3J 1P3.
An exact penalty method for solving nonlinearly constrained nonlinear least squares problems.

A E Beagles* and J R Whiteman, Institute for Numerical Mathematics, Brunel University, Kingston Lane, Uxbridge, Middlesex, UB8 3PM, UK.
Finite element treatment of singularities in elliptic problems using non-exact augmentation.

A Bellen* and M Zennaro, Institute of Mathematics, University of Trieste, 34100 Trieste, Italy.
Stability properties for continuous extensions of discrete ODE methods.

A Bjorck, Department of Mathematics, Linkoping University, S581 83 Linkoping, Sweden.
Numerical stability of the method of "Semi-Normal Equations" for solving least squares problems.

W Borchers, Fachbereich 17 Mathematik, Universität GHS Paderborn, Warburger Strasse 100, D-4790 Paderborn, W Germany.
A Fourier-spectral method for the Stokes-resolvent.

L Bos*, P Lancaster and K Salkauskas, Department of Mathematics and Statistics, University of Calgary, Calgary, Alberta, Canada T2N 1N4.
Some remarks on the representation of splines as Boolean sums.

R Bramley, Department of Mathematics, University of Illinois-Urbana, 273 Altgeld, 1409 W Green, Urbana, Illinois 61801, USA.
Constrained optimization using piecewise quadratic functions.

C G Broyden, Department of Computer Science, University of Essex, Wivenhoe Park, Colchester, CO4 3SQ, UK.
Antisymmetric matrices, staircase functions and LP.

A Buckley, Department of Mathematics, Dalhousie University, Halifax, NS B13 4HN, Canada.
Long vectors for Quasi-Newton updates.

J R Cash, Mathematics Department, Imperial College, Queens Gate, London, SW7 2BX, UK.
Extended backward differentiation formulae for stiff differential systems.

F H Chipman* and K Burrage, Department of Mathematics, Acadia University, Wolfville, NS, Canada BOP 1XO.
Linear stability of diagonally implicit general linear methods.

L Collatz, Institut für Angewandte Mathematik, University of Hamburg, Bunderstrasse 55, 2 Hamburg 13, W Germany.
Application of approximation theory to the solution of singular boundary value problems.

D Colton and P Monk*, Department of Mathematics, University of Delaware, Ewing Hall, Newark, DE 19716, USA.
Numerical results for a novel method for solving an inverse problem for time harmonic acoustic waves.

A R Conn, Department of Computer Science, University of Waterloo, Waterloo, Ontario, Canada.
On numerical methods for continuous location problems, including the handling of degeneracy.

F Corliss, Mathematics Research Center, University of Wisconsin-Madison, 610 Walnut Street, Madison, Wisconsin 53750, USA.
Adaptive self-validating numerical quadrature.

D S Daoud, Scientific Documentation Centre, P.O. Box 2441, Jadiryah, Baghdad, Iraq.
On the numerical solution of 3-dimensional linear elliptic partial differential equations.

K Dekker, Subfaculteit der Wiskunde, Rijksuniversteit te Leiden, Postbus 9512, 2300 RA Leiden, The Netherlands.
Error bounds for implicit Runge-Kutta methods in the solution of stiff nonlinear differential equations.

D B Duncan, Atomic Energy of Canada, Chalk River Nuclear Laboratories, Chalk River, Ontario, Canada K0J 1J0.
A high order leap frog scheme for hyperbolic systems.

L Elsner and M H C Paardekooper*, Department of Econometrics, Tilberg University, P.O. Box 90153, 5000 Le Tilberg, The Netherlands.
On measures of nonnormality of matrices.

D J Evans and A Danaee*, Mathematics Department, University of Isfahan, Isfahan, Iran.
A composite hopscotch method of increased accuracy.

E Eydeland, Mathematics Research Center, University of Wisconsin-Madison, 610 Walnut Street, Madison, Wisconsin 53705, USA.
Methods of computation of critical points of nonlinear functionals.

G Fairweather, Department of Mathematics, University of Kentucky, Lexington, Kentucky 40506, USA.
On the numerical solution of engineering boundary value problems.

T Fawzy, Head of Mathematics Department, Suez Canal University, 41522 Ismailia, Egypt.
Quintic spline interpolation.

J A Ford and A F Saadallah, Department of Computer Science, University of Essex, Wivenhoe Park, Colchester, Essex, UK.
A new approach to Quasi-Newton methods.

D M Gay, Bell Laboratories, 600 Mountain Avenue, Murray Hill, NJ 07974, USA.
A variant of Karmarkar's linear programming algorithm for problems in standard form.

M de Gee, Department of Mathematics, Agricultural University, De Dreijen 8, 6703 BC Wageningen, The Netherlands.
Linear multistep methods for functional differential equations.

P E Gill, W Murray, M A Saunders, J A Tomlin and M H Wright*, Department of Operations Research, Stanford University, Stanford, CA 94305, USA.
An interior-point method for linear programming and its equivalent to Karmarkar's first projective method.

N I M Gould, Department of Combinatorics and Optimization, University of Waterloo, Waterloo, Ontario, Canada N2L 3G1.
Differentiable penalty function methods for solving the nonlinear ℓ_1 problem.

W Govaerts, Seminarie voor Hogere Analyse, Rijksuniversiteit Gent, Galglaan 2, B-9000 Gent, Belgium.
The relative distance of J D Pryce in ℓ_∞^n .

I G Graham, School of Mathematics, University of Bath, Claverton Down, Bath BA2 7AY, UK.
Accurate calculation of linear functionals of solutions of certain integral equations.

J A Grant, Department of Mathematics, University of Bradford, Bradford BD7 1DP, UK.
On a method for the simultaneous determination of the zeros of a polynomial.

G Hall, Department of Mathematics, University of Manchester, Manchester, M13 9PL, UK.
Equilibrium states of Runge-Kutta schemes.

P J Harley and J D Lambert, Department of Mathematical Sciences, University of Dundee, Dundee DD1 4HN, UK.
Overlapping Runge-Kutta methods.

F K Hebeker, Fb 17 Mathematik/Informatik, Universität GHS Paderborn, Warburger Strasse 100, D-4790 Paderborn, W Germany.
Efficient boundary element methods for three-dimensional viscous flows.

B M Herbst* and J A C Weideman, Department of Applied Mathematics, University of the Orange Free State, Bloemfontein 9300, South Africa.
Recurrence in the nonlinear Schrodinger equation.

N Higham, Department of Mathematics, University of Manchester, Manchester M13 9PL, UK.
The numerical stability of two matrix Newton iterations.

N Houbak, Numerical Institute, Technical University of Denmark, Building 303, DK-2800 Lyngby, Denmark.
Improved implementation of the pivoting strategy in sparse matrix techniques.

P J van der Houwen* and B P Sommeijer, Mathematical Centre, Kruislaan 413, 1098 SJ Amsterdam, The Netherlands.
Predictor-corrector methods for periodic second-order initial-value problems.

W H Hundsdorfer*, K Burrage and J G Verwer, Mathematical Centre, Kruislaan 413, 1098 SJ Amsterdam, The Netherlands.
A study of B-convergence of Runge-Kutta methods.

R Hunt, Department of Mathematics, University of Strathclyde, Livingstone Tower, 26 Richmond Street, Glasgow, G1, UK.
The numerical solution of elliptic free boundary problems using multigrid techniques.

Z Jackiewicz, Department of Mathematical Sciences, University of Arkansas, 301 Science-Engineering Building, Fayetteville, Arkansas 72710, USA.
Boundedness of solutions of difference equations and application to numerical solution of integral equations.

R Jeltsch* and J M Smit, Institut fur Geometrie und Praktische Math, RWTH Aachen, Templergraben 55, D-5100 Aachen, W Germany.
Accuracy barriers of full discretizations of $u_t = u_x$.

I W Johnson, Mathematics Department, University of Reading, Whiteknights, PO Box No. 220, Reading, RG6 2AX, UK.
Moving finite elements for nonlinear diffusion problems.

P Keast* and M Hollosi, Department of Mathematics, Dalhousie University, Halifax, Nova Scotia, Canada B3H 4HN.
Integration formulae for singular functions over the triangle.

Ha-Jine Kimn, INRIA, Domaine de Voluceau, Roquencourt, BP 105, 78153 Le Chesnay Cedex, France.
Numerical behaviour for cubic quartic spline fits.

P E Koch, Institutt for Numerisk Matematikk, University of Trondheim, Norges Tekniske Hogskole, N 7034 Trondheim-NTH, Norway.
Multivariate trigonometric B-splines.

N Kockler, Fachbereich 17 Mathematik, Universität GHS Paderborn, Sanddornweg 37, D 4790 Paderborn-Elsen, W Germany.
Formula manipulation in numerical analysis.

A J MacLeod, Department of Management Studies, Scottish College of Textiles, Galashiels, U.K.
Parameter selection methods in the regularisation of ill-conditioned linear equations.

I Martindale* and R Wait, Department of Computational and Statistical Science, University of Liverpool, PO Box 147, Liverpool L69 3BX, UK.
Parallel algorithms for time dependent PDE's.

S P J Matthews* and R Fletcher, Department of Mathematical Sciences, University of Dundee, Dundee DD1 4HN, UK.
A variable elimination algorithm for ℓ_1 quadratic programming.

M Meneguette, Computing Laboratory, University of Oxford, 8-11 Keeble Road, Oxford OX1 3QD, UK.
Multistep multiderivative methods for Volterra integro-differential equations.

G Moore, Department of Mathematics and Statistics, Brunel University, Kingston Lane, Uxbridge, Middlesex UB8 3PH, UK.
Defect correction from a Galerkin viewpoint.

S P Nørsett, Department of Numerical Mathematics, University of Trondheim, N.T.H., N-7034 Trondheim, Norway.
Type-insensitivity for stiff codes.

G Opfer, Institut fur Angewandte Mathematik, University of Hamburg, Bundesstrasse 55, D 2000 Hamburg 13, Germany.
A "Repair" algorithm for constructing non-negative splines.

O Østerby, Computer Science Department, University of Aarhus, Ny Munkegate, DK 8000 Aarhus C, Denmark.
Step change strategies for ODE solvers.

M J D Powell, Department of Applied Mathematics and Theoretical Physics, University of Cambridge, Silver Street, Cambridge, UK.
On the numerical stability of the Bartels-Golub and Fletcher-Matthews algorithms for updating matrix factorizations.

J D Pryce, School of Mathematics, University of Bristol, University Walk, Bristol BS8 1TW, UK.
Error control of Prufer methods for Sturm-Liouville problems.

I Refat* and T Fawzy, PO Box 8878, Salmiya, Kuwait.
(0,4) Lacunary interpolation by splines.

A Salamanca* and F J Prieto, ETS Ingenerios Industriales, Jose Gutierriez Abascal 2, Madrid 28006, Spain.
On finding the minimum norm on a polytope.

S M Serbin, Department of Mathematics, University of Tennessee, Ayres Hall, Knoxville, Tennessee 37996, USA.
A new analysis of cosine schemes for second-order evolution equations.

S Skelboe, University of Copenhagen, DIKU, Sigurdsgade 41, 2200 Copenhagen N, Denmark.
Stability properties of linear multirate formulas.

D Sloan, Department of Mathematics, University of Strathclyde, 26 Richmond Street, Glasgow, G1 1XH, UK.
On non-linear instability in leap-frog finite difference schemes.

I M Snyman, Department of Mathematics, University of South Africa, Pretoria, South Africa.
Time dependent hyperbolic differential equations and variational principles.

P G Thomsen, Institute for Numerical Analysis, Technical University of Denmark, DK-2800 Lyngby, Denmark.
Runge-Kutta methods for ODE-problems with discontinuities.

J Ursell, Department of Mathematics and Statistics, Queen's University, Jeffery Hall, Kingston, Canada, K7L 3N6.
Tolerance intervals in numerical solutions of first order ordinary differential equations.

A Wathen, School of Mathematics, University of Bristol, University Walk, Bristol, UK.
Attainable eigenvalue bounds for the Galerkin mass matrix.

G A Watson, Department of Mathematical Sciences, University of Dundee, Dundee, DD1 4HN, UK.
The ℓ_1 solution of an overdetermined system of complex linear equations.

D R Westbrook, Department of Mathematics and Statistics, University of Calgary, 2500 University Drive NW, Calgary, Alberta, Canada, T2N 1N4.
Appropriate inequality constraints for contact problems of elastic beams and plates.

Xu Chengxiang, Department of Mathematical Sciences, University of Dundee, Dundee, DD1 4HN, UK.
Hybrid method for generalized nonlinear least squares.

T J Ypma, Department of Applied Mathematics, University of Witwatersrand, Johannesburg 2001, South Africa.
Efficient estimation of sparse Jacobian matrices by differences.

Y Yuan, Department of Applied Mathematics and Theoretical Physics, University of Cambridge, Silver Street, Cambridge, CB3 9EW, UK.
Computational results with a trust region algorithm for equality constrained optimization.

Z Zlatev, Riso National Laboratory, Air Pollution Laboratory, DK-4000 Roskilde, Denmark.
Variable stepsize variable formula method based on predictor-corrector schemes.